Statistics and Health Care Fraud

How to Save Billions

ASA-CRC Series on
Statistical Reasoning In Science And Society

Series Editors

Nicholas Fisher, *University of Sydney, Australia*
Nicholas Horton, *Amherst College, MA, USA*
Deborah Nolan, *University of California, Berkeley, USA*
Regina Nuzzo, *Gallaudet University, Washington, DC, USA*
David J Spiegelhalter, *University of Cambridge, UK*

Published Titles

Errors, Blunders, and Lies
How to Tell the Difference
David S. Salsburg

Visualizing Baseball
Jim Albert

Data Visualization
Charts, Maps and Interactive Graphics
Robert Grant

Improving Your NCAA˚ Bracket with Statistics
Tom Adams

Statistics and Health Care Fraud
Tahir Ekin

For more information about this series, please visit:
https://www.crcpress.com/go/asacrc

Statistics and Health Care Fraud

How to Save Billions

Tahir Ekin

CRC Press
Taylor & Francis Group
Boca Raton London New York

CRC Press is an imprint of the
Taylor & Francis Group, an **informa** business

CRC Press
Taylor & Francis Group
6000 Broken Sound Parkway NW, Suite 300
Boca Raton, FL 33487-2742

International Standard Book Number-13: 978-1-138-10639-0 (Hardback)
International Standard Book Number-13: 978-1-138-19742-8 (Paperback)

Library of Congress Cataloging-in-Publication Data

Names: Ekin, Tahir, author.
Title: Statistics and health care fraud: how to save billions/Tahir Ekin.
Description: Boca Raton: Taylor & Francis, 2019. | Includes bibliographical references and index.
Identifiers: LCCN 2018042690| ISBN 9781138197428 (pbk.: alk. paper) | ISBN 9781138106390 (hardback: alk. paper) | ISBN 9781315278254 (ebook)
Subjects: | MESH: Fraud–prevention & control | Administrative Claims, Healthcare | Statistics as Topic | Data Mining
Classification: LCC KF3605 | NLM W 86 | DDC 345.73/0263–dc23
LC record available at https://lccn.loc.gov/2018042690

Visit the Taylor & Francis Web site at
http://www.taylorandfrancis.com

and the CRC Press Web site at
http://www.crcpress.com

Printed and bound by CPI Group (UK) Ltd, Croydon, CR0 4YY

*To my mom, Dilber, and dad, Necdet,
for their unconditional love*

Contents

Preface

THE GLOBAL AVERAGE HEALTH care fraud loss is estimated to be more than 6% of the overall health care spending, or approximately more than $450 billion. In the U.S. in 2015, national health care spending was $3.2 trillion, or $9,990 per person, which accounted for 17.8% of gross domestic product (GDP) How do we ensure the fairness of all this spending? It is estimated that 3%–10% of it is lost to fraud, waste, and abuse every year.

Who has been paying for all these medical overpayments? The answer is simple. It is us, the taxpayers. The aging population increases the burden on the health care system even more. Most people believe that they will never have anything to do with health care fraud. However, all current and prospective health care users have responsibilities. Somebody out there may be using your insurance number to bill the government once every three months. You may be charged for your annual physical last year although you skipped it. Every citizen is a shareholder of the health care system.

A large number and variety of claims are processed every year. There are simply not enough experts and resources to audit every claim. New fraud types are emerging all the time. Fraudsters adapt to the preventive measures of the health care systems. These factors motivate the use of statistical and analytical methods in health care fraud detection, and hence the purpose of this book: *Statistics and Health Care Fraud: How to Save Billions*.

My interest in this topic started when I was working as a statistician for a government sub-contractor that investigates health care fraud. That helped me to understand the real nature of these problems as well as the expectations of the government and industry. The problem was way bigger and more understudied than I thought. Since then, I have been conducting academic research on this domain. My research objective is to improve health care fraud assessment by proposing methods of overpayment estimation and representative sampling as well as the use of data mining methods for fraud detection. I have written a number of peer-reviewed papers, and have given many talks to different crowds ranging from medical analysts, statisticians, computing societies, and auditors. However, I believe the major impact of this research domain will come in creating a socially significant output. In addition to the direct cost implications to the government and to the taxpayers, health care fraud also diminishes the ability of the medical systems to provide quality care to deserving patients. Therefore, I think informing the general public about the developments is very crucial. The giant problem of health care fraud can only be diminished by the help of the beneficiaries, patients themselves. The current publications about the use of statistical methods are not easy to read for people without any mathematical background. I am hoping to fill that gap and help the public to become more informed citizens.

In addition, I believe this book can help health care policymakers as well as the analysts who work in insurance programs. Select chapters can provide background about the use of statistical methods. Some chapters may be considered as complementary reading for introductory courses of applied statistics related to health care, government, accounting, and business programs. Individuals who are interested in details can check the related references at the end of the book. In case one is not interested in the content of a particular chapter at all, the overview and key takeaways can be examined for a brief understanding.

Technically, for an action to be labeled as fraudulent, the bad intent should be proven in court. The focus of this book is not on the determination of legal or medical judgments. We emphasize any kind of overpayments and respective statistical methods to reveal them. For simplicity, the term "fraud" is used while referring to any kind of overpayments including fraud, waste, and abuse.

The book is structured as follows. The first chapter presents a brief overview of global health care systems and discusses the types of medical overpayments and governmental fraud assessment initiatives using a variety of real-world examples. The second chapter introduces health care data with a discussion of data-related issues as well as descriptive statistical and visualization methods to handle health care claims data. The third chapter provides an overview of probability sampling and subsequent overpayment estimation methods in health care fraud assessment. The fourth chapter introduces data mining methods that require labeled data, such as prediction algorithms of regression and classification. The fifth chapter emphasizes the discovery of emerging and new fraud patterns. In doing so, descriptions of outlier detection, clustering and association type algorithms are provided. The book concludes with an overview of a variety of ongoing efforts, opinions, and challenges associated with health care fraud assessment, and presents a take on future developments.

Acknowledgments

Statistics and Health Care Fraud would not have been possible without the initiative of American Statistical Association and CRC Press. They have done a great job sponsoring this book series to create a forum highlighting the importance of statistical reasoning in every day life. I am beyond grateful to the series editors and the publishing team at CRC Press, especially John Kimmel. If John hadn't approached me requesting me to submit a proposal, this book would have never existed. He graciously helped with making this manuscript appropriate for the book series, offering his expertise in polishing it.

My special thanks goes to my mentor and friend, Refik Soyer. I have learned a lot from his joyful and hardworking approach to life and statistics. His vision in initiating a university-industry collaboration and our joint work presented me with the chance to get exposed to the field of statistical health care fraud assessment. My industry colleagues including but not limited to Paulo Macedo, Tim Champney, and Sewit Araia taught me a lot about health care fraud. It was a pleasure working with them. Over the years I have had the privilege to work with many bright minds to solve health care fraud problems using statistics. This list includes but not limited to Muzaffer Musal, Greg Lakomski, Fabrizio Ruggeri, Francesca Ieva, and Babak Zafari. I also want to thank Francis Mendez for giving me inspiration for being the best version of myself. I acknowledge Denise Smart and the family of Brandon

Dee Roberts for providing me with the support that enabled me to write this book.

Lastly, I want to thank my immediate family, Necdet, Dilber, Demet, Mehmet, and Kıvanç. Your support in this journey means a lot. Let me also thank my awesome wife, Bethany. From reading early drafts to giving me advice on the cover and patiently letting me work at nights, she was as important to this book getting completed as I was.

Abbreviations

ACFE	Association of Certified Fraud Examiners
AHA	American Hospital Association
AMA	American Medical Association
BETOS	Berenson-Eggers Type of Service
CDC	Centers for Disease Control and Prevention
CERT	Comprehensive Error Rate Testing
CHIP	Children's Health Insurance Program
CLT	Central Limit Theorem
CMS	Centers for Medicare & Medicaid Services
CPT	Current Procedural Terminology
DOJ	Department of Justice
FBI	Federal Bureau of Investigation
FCA	False Claims Act
FFS	Fee for Service
FPS	Fraud Prevention System
GDP	Gross Domestic Product
HCPCS	Health Care Common Procedure Coding System
HEAT	Health Care Fraud Prevention and Enforcement Action Team
HIPAA	Health Insurance Portability and Accountability Act
ICD-10	International Classification of Diseases and Related Health Problems, 10th revision
IDR	Integrated Data Repository
IRS	Internal Revenue Service
LEIE	List of Excluded Individuals/Entities

MAC	Medicare Administrative Contractors
MCO	Managed Care Organizations
MEDIC	Medicare Drug Integrity Contractor
Medi-Medi	Medicare-Medicaid Data Match Program
MSIS	Medicaid Statistical Information System
NHCAA	National Health Care Anti-Fraud Association
NHS	National Health Service
NPI	National Provider Identification
NPPES	National Plan and Provider Enumeration System
OECD	Organization for Economic Co-operation and Development
OIG	Office of Inspector General
One PI	One Program Integrity
PSC	Program Safeguard Contractor
PSR	Professional Services Review
RAC	Recovery Audit Contractor
U.K.	United Kingdom
UPIC	United Program Integrity Contractor
U.S.	United States (of America)
WHO	World Health Organization
ZPIC	Zone Program Integrity Contractor

CHAPTER 1

Health Care Systems and Fraud

OVERVIEW

"Health Care Costs Unsustainable in Advanced Economies without Reform"

(OECD, 2015)

"For the First time, Health Care Spending Higher than Social Security"

(LEONARD, 2016)

"U.S. Health Care Admin Costs Are Double the Average"

(PETROFF, 2017)

"Texas Feels Health Care's Costly Drain"

(JOSEPH AND RICE, 2017)

"U.S. Charges 412, Including Doctors, in $1.3 Billion Health Fraud"

(RUIZ, 2017)

"Bribes to Low-Paid State Worker Key to $1 Billion Miami Medicare Fraud Case, Prosecutors Say"

(WEAVER, 2017)

"Hospitals Are Clogged With Patients Struggling With Opioids"

(YIN, 2017)

Above are a number of recent headlines from official reports and newspapers. Is it surprising? Almost two decades ago, one of the rare books on health care fraud (Sparrow, 2000) started with the following: "During calendar year 2000, national health spending for the United States will exceed 1.3 trillion dollars ... This figure represents roughly 13.6 percent of the gross domestic product (GDP), up from 5.7 percent in 1965, and from 8.9 percent in 1980." At that point, the increase in the health care expenditures was a major area of concern. Fast forward to now, what has changed? This chapter provides a brief overview of the current status of health care systems with an emphasis on U.S. governmental insurance programs as well as an introduction to medical overpayments.

The increases in health care spending and health care fraud are global problems. The insurance programs in developing countries are attractive targets for fraud due to lack of monitoring. The increase in the median age and therefore in health care spending makes fraud assessment a pressing issue in developed countries. For instance, in the United States (U.S.), there have been major legislative changes including but not limited to the Affordable Health Care for America Act. However, policymakers still have not figured out how to improve health care and control spending at the same time. In 2015, in the U.S., total health care related spending grew by 5.8% to $3.2 trillion, or $9,990 per person. This accounted for 17.8% of GDP, whereas the Organisation for Economic Co-operation and Development (OECD) average was 8.9% in 2013. The U.S. spends more than twice the average health expenditure per person in OECD countries. Does this additional spending result in improved health outcomes? That is up for debate.

There are various perspectives on how to control the health care spending while keeping the quality of medical performance and coverage at high levels. Aggressive high pricing as well as a relatively low share of spending for preventive social services such as housing assistance, employment programs, disability benefits, and food security can be argued to be among the main sources of higher health care spending in the U.S. Health care fraud is not generally quoted at the top of the list. Many health care policymakers and economists do not pay enough attention to the impact of fraud, waste and abuse on the quality and cost of health care.

This can be explained by the core task of health care systems. The main focus is to deliver health care, not to carry out fraud control. Similar to any risk management framework, as long as there is no apparent fraud, the shareholders do not put much effort into investigations and they maintain the status quo. Therefore, most of the fraudsters can stay under the radar for long periods of time, and such loss remains invisible.

In the U.S., the Federal Bureau of Investigation (FBI) estimates that at least 3%–10% of health care spending is lost to fraud, waste, and abuse annually. Other developed countries are not much different. In addition to the direct cost implications, these fraudulent transactions also diminish the effectiveness of health care systems. This can partially explain the fact that the U.S. health care system does not score any better than most OECD countries with respect to many health measures, in spite of overwhelming health care spending. Can governments cope with these inefficiencies in their health care systems? Proper use of statistical fraud detection can definitely help.

In July 2017, the largest health care fraud takedown operation in U.S. history took place. More than 412 people in 41 federal districts were charged for a total of $1.3 billion in false billings to government health care providers. In addition to the financial recovery, these ongoing investigations aim to address emerging health concerns due to health care fraud. For instance, some medical facilities were suspected of preying on drug addicts. They provided unnecessary treatments in exchange for cash and drugs. These treatments worsened the condition of these people. According to the Centers for Disease Control and Prevention (CDC), every day 91 Americans die because of an opioid-related overdose. Overall, the number of prescription opioids sold in the U.S. nearly quadrupled in the last 18 years, despite the fact that there has not been an overall change in the amount of reported pain. In a *New York Times* article, the acting director of the FBI suggested that some doctors prescribe more controlled substances in one month than entire hospitals.

Increasing budget deficits and the deterioration in the quality of medical services increase the spotlight on this issue. Governmental organizations have launched a number of initiatives to control health care spending and reduce unnecessary payments.

After providing a brief overview of health care systems worldwide, this chapter presents the types of medical overpayments using a variety of real-world examples. Then, a briefing of U.S. governmental initiatives and stakeholders in health care fraud assessment is provided along with a general fraud assessment framework and a discussion of current initiatives against health care fraud.

HEALTH CARE SYSTEMS

"Everyone has the right to a standard of living adequate for the health and well-being of himself and of his family, including food, clothing, housing and medical care and necessary social services."

As stated by Article 25 of the United Nations' Universal Declaration of Human Rights in 1948, access to health care has long been argued to be a human right. Therefore, it is not surprising that most OECD countries have universal coverage for major crucial health services. While there are differences in the form of delivery between countries, health care systems are always among the largest government-led programs and have direct effects on the well-being of citizens.

How large is the health care industry? Health care expenditures constitute a significant portion of governmental budgets, especially in developed countries with high median aged populations. More than $6.5 trillion was spent worldwide in 2015. This sounds like a big amount, but what does this really mean? It can feed more than half of the world for a whole year. And this amount is not decreasing, nor even staying the same. By 2040, global spending on health is expected to increase to $18.28 trillion worldwide. Despite this, the Institute for Health Metrics and Evaluation still expects many countries to miss important health benchmarks.

In the U.S., national health care expenditures reached $3.2 trillion in 2015, both U.S. Medicare (Medicare from now on) and U.S. Medicaid (Medicaid from now on) spending more than $500 billion. This is large enough to provide median income to half of U.S. households every year. The Centers for Medicare & Medicaid Services (CMS) projects it to grow 1.2% faster than GDP per year until 2024. Projections have anticipated total health care spending to exceed $5.0 trillion in 2022. Many experts predict the bankruptcy of health care funds in the near future. Despite all this spending, U.S still lags behind most OECD countries in terms of many health indicators.

Health care systems are so giant and so variable that it is not and will not be easy to address all related issues. In the U.S., despite the lack of a universal system, governmental insurance systems still serve more than 100 million people every year. Just think about Mr. Jon Stuart, who is 67 years old and unfortunately just got cancer, or Ms. Clara Daniels, who is six months old, in acute care, and waiting for a kidney transplant, or Ms. Jane Foreman, who went to her family physician for an annual physical exam. All these people with totally different conditions may be getting service from the same governmental insurance system. For each case, some level of subjective decision-making and medical expertise are needed to choose the right treatment or prescription. That is one reason why health care systems are so complex and difficult to manage, and are attractive targets for attempts of fraud, waste, and abuse.

WORLDWIDE HEALTH CARE INSURANCE PROGRAMS

Health care insurance programs are provided by a number of distinct organizations in most countries. Now, let's present an overview of worldwide health systems with an emphasis on U.S. governmental programs.

The oldest national social health insurance system belongs to Germany, and can be dated back to Otto von Bismarck's social legislation in the 1880s. Germany has a universal multi-payer system with two main types of health insurance. Germans are offered three mandatory health benefits, which are co-financed by employer and employee: health insurance, accident insurance, and long-term care insurance. Whereas in the United Kingdom (U.K.), the National Health Service (NHS) was launched in 1948 and is free for all residents, more than 64.6 million people in the U.K. and 54.3 million people in England alone. For the 2015/16 fiscal year, the overall NHS budget was around £116.4 billion, employing more than 1.5 million people.

Most OECD countries have largely public health care systems with differences in their conduct. The French health care system provides universal coverage and is largely financed by the government. Health care in Canada is delivered through a publicly funded single payer health care system. It is mostly free and the government pays a flat rate for the services provided by private entities. In Australia, primary health care for permanent residents is funded by Australia Medicare. Residents can get free treatment in public hospitals and subsidies for treatments from health care professionals within the system.

The U.S. is the major exception among OECD countries that does not provide universal health care. Coupled with the highest per capita cost, this results in fierce debates related to the prospective health care policies of candidates every election season.

"We currently have a system for taking care of sickness. We do not have a system for enhancing and promoting health."

HILLARY R. CLINTON

"Everybody's got to be covered. ... I am going to take care of everybody. I don't care if it costs me votes or not. Everybody's going to be taken care of much better than they're taken care of now."

DONALD J. TRUMP

As candidates argued before the elections, there is lots of room for improvement in U.S. health care delivery. Most governmental health services are provided by Medicare and Medicaid. Medicare can be dated back to the Social Security Amendments of 1965 that was signed into law by President Lyndon B. Johnson. Medicare was designed to provide health benefits for all Americans over age 65. Nineteen million people signed up in the first year, and its budget was approximately $10 billion. Over time, it was expanded to include individuals under age 65 with long-term disabilities and individuals with permanent kidney failure. It is federally funded and provided health insurance to 58.5 million Americans in 2017 with a budget of $672.1 billion.

Medicare consists of four parts. Part A covers inpatient hospital stays, whereas Part B is the health care insurance that pays for some services and products not covered by Part A, generally on an outpatient basis. Part C lets private insurance companies administer Medicare benefits while Part D consists of prescription drug plans.

Medicaid is a program intended to serve mainly those with low incomes. It is jointly funded by the state and federal governments and managed by the states, which determine eligibility. Medicaid covers more than one in every five Americans. It accounts for 40% of spending on long-term care services and supports 48% of all births.

Medicaid programs display greater variability in policies and procedures than the centrally administered Medicare program. Differing policies and claims processing systems unique to each Medicaid agency make the systems vulnerable and create challenges in identifying abuse and waste. Such complexity makes administration and overseeing health care expenditures challenging. Therefore, a federal agency called the Centers for Medicare & Medicaid Services (CMS) was founded to administer the Medicare program and work with state governments to manage Medicaid.

Reimbursement for health care providers is generally either done by fee-for-service (FFS) payment system or within a managed care plan. In FFS payment frameworks, health care providers are paid a fixed rate for services they bill for. These claims are assumed to be valid unless contrary evidence is found. This framework may lead providers to bill for medically unnecessary services. As an alternative, managed care frameworks emerge as cost-efficient payment plans. They are based on contracted arrangements between managed care organizations (MCO) and government agencies. MCOs accept monthly payments for a given set of members in exchange for providing core services.

The main challenges of health care systems include managing the trade-offs between eligibility, cost, and quality. The widespread use of electronic health records and health care databases have the potential to help policymakers. In 2012, the health care system had stored roughly 500 petabytes of patient data, the equivalent of 10 billion four-drawer file cabinets full of information. The size of health care data continues to increase at an exponential rate with the growing emphasis on incorporating relevant personal data to health care systems. Such data are generated by smartphones, activity trackers, and even by running mobile applications.

Statistical and analytical methods, as well as the use of information technology, become paramount to complement human intelligence to make sense of all the data. Health care fraud assessment frameworks are no exceptions. Before presenting more detail about data in the next chapter and about analytical methods in the following chapters, first we present the nature of medical overpayments. A discussion with examples of health care fraud instances follows.

MEDICAL OVERPAYMENTS

Medical overpayments correspond to the activities of fraud, waste, and abuse in health care systems. In particular, health care fraud involves intentional deception or misrepresentation that could result in some unauthorized benefit. This benefit can be in the form of obtaining money, property, or services, avoiding payment or securing personal or business gain. For an action to be labeled as fraudulent, the bad intent should be proven in court.

Health care abuse refers to poor practices that are not consistent with benchmark medical, fiscal, or business practices, such as obtaining multiple prescriptions for the same medication from different providers. Whereas health care waste corresponds to the general waste of health care services. A doctor asking for medically unnecessary tests is an example.

Statistical methods are applicable no matter the degree of legal intent. Therefore, our focus is not on the determination of legal or medical judgments. For simplicity, the term "fraud" is used while referring to fraud, waste, or abuse, regardless of intent or necessity.

WHY HEALTH CARE FRAUD? WHY NOW?

We need to go back more than 2,300 years to revisit one of the first documented attempts of fraud. Hegestratos, a Greek sea merchant, took out an insurance policy for his old ship and its cargo. However, he had a fraudulent plan. He was planning not to carry the cargo in the first place, and sink his old boat in order to collect insurance money. His plan did not pan out as intended.

His attempt to sink his own boat was halted by the suspicious crew. Hegestratos was left in the sea by his crew and drowned. The first fraud attempt did not end well for the perpetrator.

Is the name of Charles Ponzi familiar? Mr. Ponzi raised money in the 1920s by promising clients a 50% profit within 45 days, or 100% profit within 90 days. This sounds too good to be true, does it not? He was essentially paying earlier investors using the money he raised from later investors. After a year, his so-called Ponzi scheme collapsed, costing $20 million ($255 million in 2017 dollars).

Easy money still attracts thousands of people. Bernard Madoff used a similar scheme which collapsed in 2008, costing his investors about $18 billion. This was the largest recorded financial fraud in U.S. history. And most likely it will not be the last.

In the current information and technology era, there are more tools than ever to reveal fraud. However, this works both ways; fraudsters can also come up with more sophisticated schemes. Most of you have probably been exposed to a kind of fraud attempt at some point. You may have gotten a bogus call from a call center scammer. Or it may be in the form of emails from your distant (!) relatives with the promise of retrieving a million dollars.

To address such issues, many e-commerce companies and banks have designed effective controls. For instance, your credit card may have been compromised, which prompts your bank to send you a text as an identity check. These controls are very stringent in the credit card industry, and an acceptable business risk is assumed to be as low as 0.1% of all transactions.

Whereas such loss in health care industry is estimated to be as high as 10%, which makes the health care domain an attractive area for fraudsters. Fast forward to fraud allegations and attempts in the health care industry now. For instance, you may not have heard about a recent investigation in Miami, Florida. A businessman is accused of leading a $1 billion Medicare and Medicaid fraud scheme spanning 14 years. He may face prosecution charges that he exploited his network of skilled-nursing and assisted-living facilities by filing false claims for services that were not necessary. This has been the largest single criminal health care fraud case ever brought against an individual by the U.S. Department of Justice. Why health care fraud? Why now?

Health care fraudsters are generally motivated by the large sums of money involved and the low probability of detection. The size and complexity of the systems are the main challenges for detection. Fraud is invisible by nature, and fraud examiners generally work in the dark and with much uncertainty.

Many emerging types of fraud schemes are only discovered after the fact and millions of dollars worth of overpayments. Before a substantive amount of fraud happens, the level of investment that addresses a particular concern is limited. Low investment levels result in low numbers of detected fraudulent transactions. This breeds the fraud cycle. A lack of reported fraud makes all parties content and assures them everything is under control. That results in less investment.

The resources for overseeing health care expenditures and fraud investigations are still at relatively low levels compared to other industries. For instance, *The Economist* reported that the state of New York oversaw $55 billion in annual payments and 137,000 providers with their Medicaid investigations division of 110 people in 2014.

Even when there is evidence of fraud, it is tough to measure the impact or extent of fraud. Most cases are somewhat unique. For instance, what does catching twice as much fraud really mean? Does it mean that the cases of fraud increase twofold? Or has it been the same and now you have a more reliable detection system? That is part of the uncertainty with regards to the population. Nobody knows for sure.

Another challenge is the nature of health care. The health care insurance billing systems are understandably designed to help honest providers bill easily in a transparent manner. It is assumed that honest providers are error prone. They are given notices about potential denials so they can get it right next time.

Fast processing of claims is secured by law in order to help deserving providers do their job without much financial burden. The assessment frameworks are more about correction than detection. However, this is also the Achilles' heel of the system. It lets fraudsters get large lump sums without their claims being investigated in detail.

Fraudsters are conscious and knowledgeable enough to play the system. They often get feedback through insiders, and actively test thresholds and understand rules to avoid exposure. If the claims are submitted properly, the probability of detection is very low and they will be processed and paid by the system. It may not matter if those services are really provided or not.

In addition, automated fraud control systems generally work in isolation, and not in coherence with other policing systems. Unfortunately, most risk management or detection systems are perceived within an organization in a negative manner. Other parties may think the fraud examiners are too suspicious and they need to let the system work. Large investigations and more checks risk slowing down the health care system. On the other hand, fraudulent claims are often paid out even before they are flagged for further investigation. Once fraudsters are paid, recovering that lost revenue is labor intensive, costly, and often unsuccessful.

Last but not least, the low probability of detection and conviction in addition to low severity of punishment make health care an attractive domain for fraudsters. It is generally legally difficult to make a case against health care fraud since finding evidence for documented intent can be challenging. The government's main objective, in general, is to recover its losses; therefore, it may proceed with a civil prosecution since the required burden of proof is generally at a lower standard compared to a criminal proceeding.

This low-risk, high-return nature results in health care fraud being the king of white collar crimes.

IMPACT AND IMPORTANCE OF FRAUD ASSESSMENT

The exact level of overpayments and fraud in health care is impossible to know. However, the FBI estimates it to be between 3% and 10%. The Office of Management and Budget reported that $47.9 billion was lost to fraud in Medicare alone in 2010, which corresponds to 9% of total Medicare expenditure. Whereas the Government Accountability Office reported the amount of improper health care payments to be $77.4 billion.

It is easy to be confused about the real extent when these large amounts are pronounced. We know they are large, but how large are they? How large is $50 billion? It is almost equivalent to the yearly income of one million median-wage-earning households.

The 2016 U.S. budget estimates total spending for pre-, primary, and secondary education to be $41 billion. The total loss to medical overpayments is more than the cost of keeping all schools in the nation open.

Fifty billion dollars is more than enough to feed all Americans who live in food-insecure households. Or enough to send more than five million kids to college. The sheer amount of loss is not understood properly.

This is not any different in other OECD countries. In the EU, the estimates are around 5%–6% of national health care budgets. The European Healthcare Fraud and Corruption Network estimates put the total loss in Europe at $132 billion annually.

Thanks to the continued enforcement efforts and newly enacted programs, the recovery of these losses is more successful than ever. In 2015, more than $3.5 billion, including an instance of $600 million, was recovered by settlements under the False Claims Act. The CMS revoked 28,000 provider enrollments in the Medicare program and deactivated 470,000 enrollments in 2015.

However, the magnitude of these losses is not estimated to decrease. The efforts against health care fraud are not able to change the sustained culture. We still recover only ten cents of each fraudulent dollar. Who pays for these losses?

The direct burden is shared by the taxpayers. Such wasted taxpayer money and health care resources could have been used to support the deserving providers and patients. This waste decreases access to quality health care.

Fraudulent transactions have also an adverse impact on all shareholders of the system. Insurance companies and governmental programs commit resources to chase overpayments. These losses eventually lead to increased health care costs and subsequently increased insurance premiums for consumers, a.k.a patients.

It is not only faceless money we are talking about. Health care fraud has also victims. Medical identity theft can have negative financial and health consequences on affected patients. According to Ponemon's annual study on health care identity theft, 2.3 million Americans adults were health care identity theft victims in 2014, and 65% of them had to spend 200 hours and pay $13,500 an average to resolve the crime. Patients often find that their benefits are exhausted and that they have to pay for their imposters' care or may lose access and eligibility for deserved health care.

Health care fraud can also have adverse impacts on patient health. Mistreatment or overtreatment of patients for the sake of maximizing revenue is such an example. The patients may be unnecessarily prescribed expensive but potentially dangerous drugs. Some of the prescribed drugs can be addictive, and may eventually lead to overdose use. For instance, Medicaid dental claims in Texas jumped 400% in 10 years. This increase coincides with a number of Medicaid dental clinics targeting eligible patients and overtreating their patients. In Houston, Texas, a four-year-old was one of the many cases who were overtreated. The *Houston Chronicle* reported that she had complications such as brain damage at the end of a routine appointment at a local dental clinic. Was this an isolated attempt? What other real world examples and fraud types are out there? Next, we will address these questions.

TYPES AND EXAMPLES OF HEALTH CARE FRAUD

Defining fraudulent behavior, detecting fraudulent cases, and measuring fraud losses in the health care industry are challenging. With respect to level of intent, levels of overpayments can be listed as simple mistakes, inefficiencies, bending the rules, and intentional deceptions. The levels and types of fraud, waste, and abuse vary.

Fraudsters are long assumed to either play the short game with a "hit and run" strategy or the long game by using a "steal a little, all the time" scheme. The short-term strategy involves billing for large amounts generally using stolen patient IDs and then disappearing even before the fraud is revealed. In general, this is committed by organized crime networks that bill from bogus locations using false identities. Such professional fraudsters are knowledgeable about fraud detection systems and may even have access to insider information. These fraud schemes eventually get discovered and the fraudsters are suspended. However, most of the time they profit a lot in the relatively short time of their operation before getting caught.

The long-term strategy is based on stealing a little, often just below the thresholds of the automated systems. The services billed by fraudsters are mostly legitimate, therefore they are not suspected to involve fraud at all. Even if they are caught, it would mostly be assumed to be a clerical error and they would just be asked to correct for the incorrect billings.

One such example is an individual without any criminal history lending his patient ID to fraudsters and letting it be charged at certain intervals for services that he/she does not need or receive. If the dollar amount per incident is relatively low, this can easily stay under the radar. Another example can be a doctor/nurse/billing specialist who charges Medicare for a more expensive service each time. As long as these do not become excessive and are within the allowed billing norms, it would be very difficult for the detection systems to reveal the fraud. Those small amounts add up to become profitable in the long run.

Fraud can also be classified with respect to who commits it: provider, patient, and insurer. Improper billing, receipt of kickbacks, and self-referrals can be labeled as provider fraud. Patients can commit fraud by falsifying documents or misusing their insurance cards. Whereas insurers may not provide some of the services they collect premiums for. A number of typical schemes along with real-world examples follows.

Identity theft and fraud: Identity theft can happen in a variety of ways. In order to process a legitimate claim to be paid by the system, the medical history and identification card of the beneficiary are mostly enough for fraudsters.

Identity theft is often an issue with very old beneficiaries. The ID owner may be totally unaware of such fraudulent billing. For instance, a revealed fraud scheme included suspects billing for bogus ear care procedures purportedly done in nursing home facilities. Some of the older patients were even unconscious. In some cases, beneficiaries may collaborate with the fraudsters. Some health care providers were documented to offer kickbacks to patient recruiters to help assemble bogus patient information. Patients themselves may also misrepresent information about their situation during a visit.

Identity fraud corresponds to unlawful activities that use the identity of another person fraudulently. This can involve the use of falsified documents such as fake ID cards. Patients may use another person's coverage or insurance card to illegally claim the insurance benefits. Insurance subscribers can falsify employment and eligibility records to obtain lower premiums or government subsidies. For instance, they can knowingly claim an undeserved exemption from prescription charges or can claim unpaid coverage.

Improper coding: Improper coding is one of the major ways to commit health care fraud. It corresponds to incorrect billings where the billed procedures do not match the actual procedure provided. This can happen as part of clerical mistakes or deliberate attempts to increase revenue.

In some cases, improper coding is revealed when one doctor's billing activity is drastically different than his/her peers. For instance, Dr. Jacques Roy from Rockwall, Texas, was found to sign off on more home health services through Medicare than any other medical practice in the U.S. between 2006 and 2011. His billings cost the Medicare system more than $350 million. His lawyer suggested Dr. Roy had been very hardworking. The jury did not agree. After 22 days of the trial in a Dallas federal courtroom, he and his fellow perpetrators were given stiff penalties and sentences. It took investigators six years to detect this fraud. Maybe, the greed and confidence of being undetected made the fraudulent amount as large as $350 million, and this became too big to miss. There will be more on Dr. Roy in the following chapters.

Major types of improper coding can be listed as upcoding, unbundling, multiple billing, and phantom billing.

Upcoding refers to billing for a more expensive service or procedure than the one actually performed. Misrepresentation of uncovered treatments as medically necessary covered treatments to obtain payments is one example. For instance, the cosmetic surgery procedures such as "nose jobs" are not usually covered by insurance plans, therefore they are billed as deviated septum repairs. There are also cases of conspiracy fraud which require the collaboration of provider and beneficiary to commit fraud. Some physicians are found to waive patient co-pays or deductibles for medical or dental care and then later, over-bill insurance carriers or benefit plans.

Unbundling is creating separate claims for services or supplies that should have been grouped together. For instance, a provider may bill each stage of a procedure as if they were separate procedures with the goal of collecting more money from the insurance company.

Providers may also submit the same claim multiple times (twice) for the same service, in order to get paid more than once for performing the same action. Another form of multiple billing is committed by fraudulent networks. They use a single patient ID to generate billings across multiple providers. Fraudsters may get away with this since automatic processing of claims is mostly designed to improve processing speed, often without a comprehensive check.

Phantom (ghost) billing corresponds to billing for health care services that have not been provided or for medicines/medical devices that have not been delivered. Genuine information, which may be obtained through identity theft, can be used to process such claims correctly. Patient transportation services claiming charges for patients who were never moved or submitting durable medical equipment claims for services and supplies not provided are examples. Using the IDs of ghost employees/deceased employees, or submitting claims for deceased/ineligible members are major ways of committing this crime.

Other cases of phantom fraud involve the fraudsters recruiting patients for bogus procedures, doctor and pharmacy shopping, and sometimes trading narcotics in exchange for member IDs. For instance, a gang of fraudulent medical-supply stores hired recruiters to bring them patients who are eligible for government services. Then they paid off local doctors to prescribe these patients expensive motorized wheelchairs but instead delivered cheaper motor scooters, pocketing the difference. Any ideas about the extent of their fraud?

Sixty-one percent and $95 million… These were the percentages of Medicare claims for medically unnecessary power wheelchairs and the respective annual loss in the 2010s. In an investigation, the Office of Inspector General (OIG) found that 80% of claims for power wheelchairs should not have been paid at all since they did not meet the coverage requirements.

Luant & Odera, Inc., which was already billing with stolen beneficiary identities, got more creative in the aftermath of Hurricane

Katrina. It billed Medicare under a special code that designated power wheelchairs as replacements for wheelchairs lost during Hurricane Katrina when in fact the hurricane did not damage the wheelchair, or the beneficiary did not even have a power wheelchair to begin with.

In the case of Cooper Medical Supply, the owner used fraudulent prescriptions and medical documents to submit false claims to Medicare for expensive, high-end power wheelchairs. This was only revealed because more than 80% of the beneficiaries lived over 160 kilometers (approximately 100 miles) away from Cooper Medical Supply.

In another scheme, Positive Home Oxygen recruited a doctor to sign certificates of medical necessity for power wheelchairs for patients who did not meet the coverage requirements. All these fraudulent cases resulted in federal prosecution, but there are potentially many out there that go unnoticed. Despite all the attention, fraud continues to evolve and exist. For instance, a few years after all the attention that power wheelchair companies have gotten, another company was found to over-bill Medicare by as much as $108 million in a four year period.

Kickback schemes and self-referrals: Kickbacks correspond to cases of unnecessary or illegitimate health care services provided with knowledge of financial gain. For instance, pharmacists can choose to fill a prescription for drugs of a particular brand. In return, they get payments from that pharmaceutical company. This fraud scheme of Medicare Part D can also have adverse effects on patients' health. Pharmacists can collaborate with pharmaceutical companies, providers of medical equipment or health services in their attempt to manipulate prices, and maximize their revenue in cases of varying rates for services.

Some physicians may refer the patients to another provider not out of the medical necessity, but rather for financial gain. Such relationships may operate within a kickback or commission based scheme. For instance, a provider may receive cash or below fair market value rent for medical office space in exchange for referrals. This can be especially a concern for large hospital networks. Medical expertise is required to determine the legitimacy of referrals. It is generally difficult, time-consuming, and costly to decide whether the self-referrals within the same network are of necessity or not.

Incidents of referrals to "phantom companies" are also among such fraud types. For instance, in New Orleans, two doctors and a registered nurse created a six-year scheme in that they referred patients to four phantom companies for home health services and treatment. In some cases, the treatments were not even provided. The scheme was estimated to cost Medicare $50 million. The perpetrators were given prison sentences and monetary fines in federal court in 2015.

Providing unnecessary care: In these cases, physicians may perform medically unnecessary services simply to generate insurance payments. Especially in the specific medical domains where subjectivity reigns, the abuse of unconditional trust in physicians can prove to be a major problem. It may be difficult to judge the appropriate level of health care to heal the patient. The major examples of such fraud are falsifying diagnoses to justify medically unnecessary procedures, requesting unnecessary blood tests and MRIs, billing cosmetic surgeries as necessary repairs, home health care companies, and visiting nurses billing additional amounts. The fee-for-service model may motivate the physicians to maximize the number of services they bill in order to maximize their revenue.

In some of the related investigations, companies end up settling with the government without awaiting the conclusion of the audit. For instance, 21st Century Oncology was accused of performing a particular bladder-cancer test more often than medically necessary. *The Wall Street Journal* reported that it agreed to pay $19.75 million in settlement while neither admitting nor denying any wrongdoing. Settlements may be preferred especially by the large hospital networks in order to avoid bad press and the risk of being prosecuted for wrongdoing if they are found guilty.

Providing incorrect care or manipulating billing rules: It is not rare that patients do not have comprehensive coverage and lack access to the prescription they need. Some providers may want to help these patients. However, deliberately using a wrong diagnosis to be able to prescribe certain drugs to a patient is deemed as fraudulent. Such fraud may be done in good faith and doctors' perception that the health care system does not work to the full benefit of the patient. In order to provide the needed but uncovered care, the physician may manipulate the billing rules. A national survey of physicians in 1998 found that 39% reported manipulating reimbursement rules at least once so patients can receive medically necessary care.

This brings up the debate of the extent of physician independence in billing. The billings involve subjective judgments, and in some cases there are legitimate differences of opinions between doctors. In these cases, while doctors argue for the medical necessity of the billing, prosecutors may provide evidence of differences from generally accepted physician billings. Are these billings a result of aggressive prescriptions or malpractice?

For instance, a physician in Florida billed $8.4 million for laser treatments compared to average billing of $6,061 by similar doctors during a six-year period. Another doctor treated patients with an eye injection more than others, billing this procedure for $57.3 million versus the $3 million average billing of other eye doctors. He argued for the medical necessity; however, the prosecutors suggest his billing pattern made him as much as $72,000 in annual profit for each patient compared to his peers. The prosecution argued that he falsified the initial diagnosis and gave patients multiple unnecessary injections. His lawyers denied any wrongdoing, and suggested that "There is an old saying that when you look at the world through a dirty window, everything looks dirty." The case is still going on, while this physician could face up to 610 years in prison.

Managed care fraud: Managed care has emerged as a cost-efficient alternative to the fee-based payment systems. It is based on physicians being reimbursed a flat rate for patients in their area. However, it is not immune from fraud and abuse. Managed care organizations may be involved with denial or late delivery of services to deserving patients, providing substandard care, and creating logistical and administrative obstacles for patients. Billing a patient more than the co-payment amount for covered and prepaid services is an example of such fraud. In a report, the OIG outlined that more than half of providers did not even offer appointments to enrollees while 35% of providers could not even be found at the location listed by their plan.

Prescription fraud and drug abuse: Prescription fraud, waste, and abuse can occur through illegitimate, excessive, or unnecessary prescriptions. For example, some beneficiaries are found to be involved in doctor shopping which involves obtaining high amounts of prescriptions from multiple prescribers and pharmacies. In so-called drug diversion, the beneficiary or the provider can redirect prescription drugs for illegal gain. This can also take place as part of a network that involves kickback payments for writing prescriptions or dispensing drugs. Some patients are found to resell the drugs on the street. A number of schemes include physicians prescribing opioids to complicit patients that received cash for their participation in the scheme.

Opioid abuse is one of the emerging concerns, which can result in adverse effects to patient health in addition to its cost. The amount of prescription opioids sold in the U.S. has nearly quadrupled in the last 18 years. Despite this increase, there has not been an overall change in the amount of reported pain. According to the Centers for Disease Control and Prevention (CDC), every day 91 Americans die because of an opioid-related overdose. The acting director of the FBI suggested that some doctors wrote out more prescriptions for controlled substances in one month than entire hospitals were writing. It has reached such high levels that the President declared a national opioid emergency. As part of these health care fraud enforcement actions, licensed medical professionals across 41 federal districts were charged for their alleged participation in health care fraud schemes involving approximately $1.3 billion in false billings. In July 2017, this was also reported as the "largest opioid-related fraud takedown in history," including doctors allegedly running pill mills and operators of fraudulent treatment centers.

These are the main types of health care fraud along with some examples. What do officials do to control the extent of health care fraud, waste, and abuse? Next, we present the frameworks and initiatives to address these overpayments.

GENERAL FRAUD ASSESSMENT FRAMEWORK AND INITIATIVES

The majority of the health care claims are submitted electronically and processed automatically by computerized, rule-based systems. Health care anti-fraud efforts rely on a combination of pre-payment edits and audits as well as post-payment utilization review and special investigations. The Office of Inspector General (OIG) and the Centers for Medicare & Medicaid Services (CMS) are the core agencies responsible for overseeing the integrity of the Medicare program, and sharing oversight of Medicaid programs with other state-level authorities. In order to support its efforts to prevent, detect, and investigate potential health care fraud and abuse, the CMS also works with an array of contractors.

Generally, Medicare Administrative Contractors (MACs) conduct the initial edits to evaluate the eligibility of claims with respect to the Health Insurance Portability and Accountability Act (HIPAA) standards and CMS guidelines. MACs are private health care insurers that have been awarded a geographic jurisdiction to process Medicare Part A and B health care claims or durable medical equipment claims for Medicare FFS beneficiaries. If errors are detected at this level, the entire batch of claims would be rejected for correction and resubmission.

Claims that pass these initial audits are then checked against implementation guide requirements such as the coverage and payment policy. In case of error detection at this level, only the individual claims that included those errors would be rejected for correction and resubmission. If the claims satisfy the criteria of the edits and audits built into the system, then automatic payment follows, generally without any human involvement. Most paid claims are not subject to any human scrutiny.

The payment systems focus on processing accuracy and fast processing, but generally without verifying the legitimacy of the claims. In the past, the OIG conducted "Medicare Overpayment Rate" studies to assess the accuracy of these payment systems. They conducted post-payment audits of statistically valid samples of paid claims. The overpayment rate was found to be as high as 14% in 1997 and decreased to almost 7% by 2002. However, these only measured the efficiency of the payment system, assuming that all claims were billed for medically necessary and provided services.

Medical review audits are based on the assumption that submitted claims are billed for services that are actually provided and medically necessary based on correct diagnoses. All participants of the system are assumed to have good intent and appropriate medical judgment, and may make occasional processing mistakes. The system is designed to catch those mistakes and provide feedback to ensure they do not happen again. They prevent payments for uncovered, incorrectly coded, or inappropriately billed services. This generally does not involve criminal control.

However, in the case of dealing with fraudsters, such a claims-processing process is vulnerable. It can actually help fraudsters learn about the deficiencies and limitations of the systems and improve their fraud scheme.

Having seen these issues, the legislators funded Medicare via the 2010 Small Business Jobs Act to apply analytical methods to prevent improper payments. Existing medical review audits are supported with pre-payment reviews using analytical method-based Fraud Prevention System (FPS). The CMS announced that it had prevented $42 billion of improper payments to health care providers in fiscal years of 2013 and 2014. Most of the savings were the result of prevention activities. FPS functions within the congressionally mandated Medicare payment window of 14 to 30 days, preventing payment delays to legitimate practitioners.

Despite these recent initiatives to strengthen pre-payment audits, anti-fraud efforts are traditionally based on so-called pay-and-chase methods. These mainly include post-payment utilization review and special investigations. In general, a small proportion of paid claims are selected, and health care providers are asked to provide further documentation of the relevant medical records in typically 90 days. Once received, these medical records are reviewed and compared with the claims they are supposed to justify. Post-payment audits focus on medical appropriateness, not truthfulness. Therefore, this step may not be efficient, especially if fraudsters lie twice by fabricating matching records that correspond to their fraudulent claims.

In addition, special health care investigations are conducted if there is enough suspicion and evidence of wrongdoing. The goal is to identify and correct improper payments through efficient detection procedures. Each health care investigation requires the subject domain expertise of licensed professionals that manually audit claims. Such audits are costly, but the overpayments from a smaller group, aka sample, of claims can be extrapolated to the overall set of claims under certain conditions. The trade-off between the audit costs and accuracy of the extrapolations is one of the main challenges in subsequent resource-allocation decisions. The audit costs correspond to the time spent by the expert as well as the physical resources. The unnecessary audits of claims that are indeed legitimate result in loss of trust in the government and lost opportunity cost.

Other contractors include Comprehensive Error Rate Testing (CERT) Contractors, Medicare Drug Integrity Contractors (MEDICs), Zone Program Integrity Contractors (ZPICs), and Recovery Audit Contractors (RACs). CERTs help calculate the Medicare Fee-for-Service improper payment rate by reviewing claims to determine if they were paid properly. MEDICs monitor fraud, waste, and abuse in the Medicare Part C and D programs. RACs reduce improper payments by detecting and collecting overpayments and identifying underpayments. ZPICs, formerly called Program Safeguard Contractors (PSCs) investigate potential fraud, waste, and abuse for Medicare Part A and B in addition to the areas of durable medical equipment, and home health and hospice. ZPICs investigate leads through boots-on-the-ground activities such as site visits, beneficiary interviews, and medical chart reviews.

There are a number of major federal laws that address the cases of revealed health care fraud. The False Claims Act (FCA) prohibits people from knowingly submitting claims involving some material misstatement or deception. Any kind of overpayment should be returned to the system within 60 days; otherwise, the provider may be liable for fraud. For instance, upcoding may result in fines of up to three times the amount of damages sustained by the government and up to $21,563 (in 2016) per false claim filed. Most of the false claims are revealed through the cooperation of whistleblowers who used to be employees at the company and have a good understanding to report the business policies. The FCA includes protections for whistleblowers in its qui tam provisos.

The Anti-Kickback Statute addresses the issue of kickback payments, and its violation can result in fines up to $73,588 (in 2016) per kickback plus three times the amount of the kickback. The Physician Self-Referral Act (Stark Law) was enacted to address physician self-referral in cases such as referrals to clinical laboratories, multitude of diagnostic and treatment self-referrals of Medicare patients by physicians or their family members.

The Civil Monetary Penalty Statute was enacted to protect the well-being of patients. It prohibits any entity from filing a medical claim for a service not provided as claimed, that is fraudulent, or that is not medically necessary, or when such a payment provided a physician with any incentive or payment that could possibly result in decreased care of a patient. The Criminal Health Care Fraud

Statute prohibits knowingly and willfully executing or attempting to execute an unlawful coordinated scheme that defrauds any health care program. The OIG has the right to exclude fraudulent providers from participation in all health care programs.

If any evidence of wrongdoing is found, the CMS may take a number of actions, including pre-payment review, suspension of payments, and termination of a provider's billing privileges. Such cases also can be referred to law enforcement. After exhausting internal and qualified independent contractor appeals, the providers have the right to take CMS decisions to judges and appellate courts including the Administrative Law Judges, the Medicare Appeals Council, the Department of Health and Human Services Appeals Board, and the Provider Reimbursement Review Board. If these appeals do not satisfy the provider, the decision can be taken to the federal district court.

Fraud detection efforts are very crucial, because they can help change the culture and inhibit the growing tendency for criminal activities. It is widely estimated that federal health care spending will increase, so the government needs to find solutions that decrease the momentum of fraudsters. Hence, fraud assessment and detection are necessary for any health care insurance program. In the following chapters, statistical methods that enable auditors in this battle are presented.

KEY TAKEAWAYS

1. Health care expenditures constitute a significant portion of governmental budgets, with a total of $6.5 trillion spent worldwide in 2015.

2. Medical overpayments correspond to activities of fraud, waste, and abuse, which differ in the level of intent and necessity.

3. The proportion of medical overpayments in the U.S. is estimated to be at least 3% of overall expenditures.

4. Health care fraudsters are generally motivated by the large sums of money involved and the low probability of detection due to the size and complexity of the systems.

5. The main types of fraud can include identity fraud, improper coding (upcoding, unbundling, multiple billing, and phantom billing), kickback schemes, and self-referrals, providing unnecessary care, providing incorrect care, or manipulating billing rules, managed care fraud, prescription fraud, and drug abuse.

6. In the U.S., the OIG and CMS oversee the integrity of the health care programs with respect to federal laws such as The False Claims Act.

ADDITIONAL RESOURCES

1. Health care fraud, definitions, examples, and resources:

   CMS (2016). Medicare fraud and abuse. The Centers for Medicare & Medicaid Services. https://go.cms.gov/VCT5iH

2. Health care systems statistics:

   CMS (2017). NHE fact sheet. The Centers for Medicare & Medicaid Services. https://go.cms.gov/1O1cWhR

3. Publicly available fraud and abuse products:

   CMS (2018). Medicare Learning Network Fraud and Abuse Products. The Centers for Medicare & Medicaid Services. https://www.cms.gov/Outreach-and-Education/ Medicare-Learning-Network-MLN/MLNProducts/Downl oads/Fraud-Abuse-Products.pdf

4. Opioid overdose statistics:

   CDC (2016). Drug overdose deaths in the United States continue to increase in 2015. Centers for Disease Control and Prevention. https://www.cdc.gov/drugoverdose/epidemi c/index.html.

5. Scholarly papers with comprehensive literature review of health care fraud assessment:

- Ekin, T., Ieva, F., Ruggeri, F. and Soyer, R. (2018). Statistical medical fraud assessment: Exposition to an emerging field. *International Statistical Review.* 86(3), 379–402.

- Li, J., Huang, K.-Y., Jin, J. and Shi, J. (2008). A survey on statistical methods for health care fraud detection. *Health Care Management Science*, 11:275–287.

CHAPTER 2

Describing Health Care Claims Data

OVERVIEW

Health care data grow rapidly in volume, variety, velocity, veracity, and value. This chapter introduces health care data with an emphasis on the related challenges in health care fraud assessment. Standard health care data types are in the form of practitioners, clinical instances, and claims data. In health care fraud investigations, complementary information from sources such as public records, social media, and weather reports can also be used. These additional data sources include driver's licenses, birth/death, voter, banking, property tax, and commercial records.

The incomplete, subjective, and dynamic nature of health care data is among the primary challenges associated with health care fraud data assessment. As in any other data analysis procedure, the most important aspects include integrity, pre-processing, transformation, storage, and security of data. Data integrity and pre-processing are very crucial since the quality and validity of the data used for analysis form the basis for all results. Data pre-processing takes most of the time in many data analysis projects. This is especially true with health care data because of their variable nature. Data security and proper access are other important aspects. There is a significant amount of confidential data in health care. Access is regulated by a number of federal acts including the Health Insurance Portability and Accountability Act (HIPAA). For compliance, it is important that only authorized people can access the data of concern.

Tools of descriptive statistics to visualize the data can help detect obvious patterns and reveal potential cases of overpayment, especially in cases with a small number of claims and in apparent overpayment cases. For instance, such descriptive methods are used to identify excessive charges outside the thresholds. Visualization-based tools enable the auditors to reveal excluded and invalid providers, vendors, and patients. Then auditing can be conducted on the claims of interest. Even automated rules can be created to deal with these excess activities, and can prevent payment of these claims in the first place.

This chapter focuses on understanding health care claims data as well as types of data pre-processing. Descriptive statistical analysis and visualization methods are described. These are powerful on their own in addition to being used complementarily to many statistical methods in health care fraud assessment. For instance, after grouping the claims, descriptive statistical summaries of the variable of interest can help to understand the differences among groups. In the case of grouping, the mean and standard deviation of the audited claims in each subgroup can provide evidence about the success of the method of choice. The importance and use of descriptive statistical summaries are presented throughout this book.

HEALTH CARE DATA

The types of health care data vary from program to program. It can correspond to electronic health and medical records, prescription, lab tests, patient registry, and payer claims. Standard health care data are generally in the form of practitioners' data, clinical instance data, or health care claims data. Practitioners' data summarize the activities of service providers in a certain time period and list provider related aspects such as service cost, usage, and quality. Clinical instance data consist of activities performed by medical staff in a particular treatment.

Most raw health care data are in the form of claims. A health care claim involves the participation of a patient and a service provider, and contains the attributes of patients, providers, and the claim itself. Claims generally include the patient name and ID, provider name and ID, date of service, diagnosis, type of procedure, billed and paid amounts, diagnosis outcome, and prescription details. Additional attributes of a patient can be gender, location, age, and medical history, whereas the type and the location of the facility are among the attributes of a provider.

Practitioners' data, such as provider practice information, could be augmented with claims to verify addresses and the validity of a current provider practice. Furthermore, one can aggregate provider data to identify excess activities such as overcharging or overtreatments. Such provider variables include

- Total amount billed

- Total number of patients

- Total number of patient visits

- Per-patient average amounts of payments/billings

- Per-patient average number of visits

- Per-patient average number of medical tests

- Per-patient average costs of medical tests

These statistics can be used to detect irregular activities. For instance, CNN reported that four people were charged for sending talking glucose monitors to Medicare patients across the country who did not need or request them. They received more than $22 million from Medicare, which was way higher than the comparable activity of relevant providers. Excess activity in referrals also can indicate wrongdoing.

A urologist in Florida was found to be the number one referring physician in the U.S. with respect to fluorescence in situ hybridization (FISH) tests. Later, a Department of Justice (DOJ) investigation revealed that he was paid approximately $2 million in bonuses based on the number of tests he referred to. He eventually agreed to pay back $3.81 million as part of a settlement. That lab company also settled for a $19.75 million agreement to resolve these allegations.

Providers can argue that their billings are higher since they are taking care of more sick and older patients. To address that concern, auditors use health risk scores of patients to assess the level of sickness and risk. For instance, while comparing the billings for a particular diagnosis, one should compare providers that serve patients with similar health profiles.

The governmental organizations have become more transparent and have made more health care claims data publicly available. In 2014, the Centers for Medicare & Medicaid Services (CMS) released a public data set of the Medicare Part B insurance program. This data set includes a set of records documenting information about transactions between over 880,000 health care providers and the CMS, totaling over $77 billion of payments for the year 2012. As of now, the Medicare claim payments between 2013 and 2016 are available in the Research, Statistics, Data & Systems page of the CMS data website (CMS, 2016a).

Another publicly available data source is the List of Excluded Individuals/Entities (LEIE) data set (OIG, 2017), which is updated regularly by OIG. This data set is useful to ensure these excluded providers are not paid for any claims they may submit. The Office of Foreign Assets Control's Specially Designated Nationals and Blocked Persons list can help revealing collusion with foreign entities. ProPublica, a nonprofit that provides investigative journalism, can be relevant to access the information about providers under investigation.

Information from sources such as public records, web searches, social media, and weather reports can also be used. These public records include driver's licenses, birth/death, voter, banking, property tax, and commercial records. Such records and social media can be used to reveal relationships to potentially understand fraudulent networks. For instance, information that is available at the website of a provider can help understand his/her office hours. If a provider has unusually high billing activity for a Sunday when he/she does not hold office hours, that may raise a flag.

Weather events can be taken advantage of as well. For instance, the day after Hurricane Harvey hit Houston, if a provider has billed for 40 Medicare patients in a flooded area, that can be flagged easily for further investigation. This basically means that 40 mostly senior patients reached the provider's office battling five inches of rain, as if it was a regular day.

In addition to these standard sources, patient lifestyle information can be used complementarily. Many patients generate data themselves through mobile applications for personal fitness or activity tracking devices. These have the potential to improve personalized medicine, which can enable fraud investigators to measure patient risk profiles more accurately.

UNDERSTANDING HEALTH CARE CLAIMS DATA

A standard web query for health care claims data returns more than 500 million results. CMS health care claims data sources are among the prominent public data sources. Most of the time, public health care claims data are presented after aggregation is done for each provider. The beneficiary information is de-identified so that the patient characteristics are kept private. In health care claims data, some of the widely used variable descriptions can be listed as:

- National Provider Identification (NPI): an identifier for each health care provider based on the National Plan and Provider Enumeration System (NPPES)

- Specialty: the medical specialty of the provider, such as pediatrics, cardiology, internal medicine, or diagnostic radiology

- Health care Common Procedure Coding System (HCPCS) code: the identifier of each medical procedure or service. HCPCS includes three levels of codes:

  - Level I: consists of American Medical Association (AMA) Current Procedural Terminology (CPT) codes

  - Level II: HCPCS alphanumeric code set and primarily includes non-physician products, supplies, and procedures not included in CPT

  - Level III: HCPCS local codes, developed by state Medicaid agencies, Medicare contractors, and private insurers for use in specific programs and jurisdictions

- Beneficiary count: the number of distinct Medicare beneficiaries per day that are served by a particular provider

- Service count: the number of health care procedures per day that are billed by a particular provider

Each claim also has the name, gender, and credentials of the physician, the entity code, address of the provider, place of service, procedure code description, and modifiers. The mean (average) and standard deviation (a variability measure) of the submitted amount and the Medicare allowed amount for that particular procedure are also generally listed. More information on these descriptive statistical measures is provided later in this chapter.

DATA PRE-PROCESSING

Data integrity is one of the most crucial components of any type of data analysis. In information systems, there is a saying: garbage in–garbage out. Making decisions based on invalid data could result in severely negative impacts. You cannot do proper statistical analysis without high-quality data. What is really meant by that?

There may be many reasons for a decrease in the quality of the data. Data points may not be recorded into the system correctly in the first place. Or you may have a variable that has an abundance of missing values. Such missing data can produce problems by telling only portions of the whole story.

A fundamental question in data analysis is how to deal with missing data. I am sure one of the first things that comes to your mind is simply removing the claim lines with missing information especially if you have a large data set. And you are not alone. That is one of the most popular and straightforward methods. However, keep in mind that removal of instances can decrease the statistical power of an analysis, since potentially valuable information in the other fields is also lost. You may also create a so-called biased data set that emphasizes the records without missing information. Let me present this with a hypothetical data set, see Table 2.1.

Let's suppose Dr. Groux has billed for five patients. This is a data set of only five data points that have missing information. If you remove the third observation since it is missing the payment amount, you would end up with only four. You may lose valuable information since that is the only claim that has a 40-year-old patient. Therefore, that may not be desirable. Most of the time, we do not work with such small data sets. However, the idea of handling missing data is similar. So, what do we do?

TABLE 2.1   Billing Data of Dr. Groux

Index	HCPCS Code	Description	Age	Payment Amount
1	92004	Eye exam–new patient	50	50
2	92014	Eye exam–treatment	60	50
3	92014	Eye exam–treatment	40	?
4	92014	Eye exam–treatment	20	30
5	92081	Visual field examination(s)	20	10

Popular options include replacing missing values with the most likely value or the average of the data set or even user-defined constants. So, what are the average and most likely payment values? Would you compute the average for the whole data set or only the related procedure? These generally have open-ended responses. That is why it is important to work with domain experts.

An alternative is to decide with respect to the payment amounts in the data set. For instance, in this data set their values are 50, 50, 30, and 10. One can argue that the probability of the next observation being 50 is two out of four, or 50%. More advanced statistical analysis methods can be used to predict the missing values—so-called imputation. The predictive methods will be presented in Chapter 4.

Another important step in data analysis is data pre-processing and transformation. Eighty percent of the time for analytical projects in the industry is estimated to be spent on data pre-processing efforts such as data cleaning and transformation. Health care fraud assessment is not an exception, and it has unique challenges. For instance, providers can submit the same claim using the name of the hospital or the provider identification number. Therefore, most analysts end up defining new unique identifiers to analyze health care claims data. For large data sets, there is the issue of variables with blank records for some observations.

Choice of variables of interest depends on the objective of a particular analysis. For instance, analysts may prefer to check out the frequent fraud occurrences or fraud types with the most financial losses. Fraud investigators also refer to this as "following the money," since they prefer to focus on moving high dollar cases.

The variables about the identity of the claim, provider, and patient, clinical insights, monetary amount, and the paid portion for that claim are generally considered. The financial information of the provider and patient history would also be helpful depending on the particular case.

One of the biggest challenges is the lack of data labeled as fraudulent. Labeling data requires an actual audit, therefore it is time-consuming and expensive. Fraud is deceptive in nature, and it is a relatively rare event. These characteristics correspond to a lack of information and labeled data. Another issue is the dynamic nature of fraud patterns because fraudsters change their tactics often to navigate away from audits in a particular area. This is a constantly changing battle. Even legitimate claims have changing patterns due to heavy competition in the health care industry and the changing legal framework. And keep in mind that traces of fraud are investigated in a world of many prescriptions, provider types, and various patient needs. Where do we store all these large data sets?

Medicare claims data are stored in the continuously expanding Integrated Data Repository (IDR) since 2006. This database provides a comprehensive view of data, including claims, beneficiaries, and Part D drug information. Such a comprehensive look is especially beneficial for pre-payment review tools such as identity checks. These can be conducted by simple comparisons of data from different sources and can be very helpful in pre-payment analytics. The IDR can also be accessed through the centralized, web-based One Program Integrity (One PI) Portal. One PI caters to in-house CMS specialists, supporting contractors, and law enforcement. The analytical tools such as the Fraud Prevention System pull program data from these sources in order to leverage sophisticated methods.

Whereas, for Medicaid and the Children's Health Insurance Program (CHIP), there is the national database of the Medicaid Statistical Information System (MSIS). It helps states and the federal government to gather key eligibility, enrollment, program, utilization, and expenditure data from all states, in addition to the particular repositories of the states. The computations generally involve intensive tasks. Therefore, it is common to use distributed storage and computing tools.

Comprehensive access to state repositories is given after careful authorization checks and for particular audits. Data security and compliance with data access protocols are very important.

All shareholders of the system need to comply with federal acts such as HIPAA and the Federal Information Security Management Act.

DESCRIPTIVE STATISTICAL ANALYSIS

The health care fraud assessment industry has started to employ more sophisticated methods in recent years. However, in reality, most of the detection so far has been made by less exciting but very beneficial, relatively simple methods. Many existing fraud detection solutions are based on a set of business rules that detect suspicious or anomalous payment events. For example, a rule might look for payments for a given medical procedure where the amount is considerably higher than the average payment for that procedure. In the following, such descriptive statistical summary measures and related visualization tools are presented. These can help with:

- Analysis of a large amount of information

- Detection of general patterns and trends

- Detection of outliers and unusual patterns that can indicate fraud

- Identification of potential mismatches and frequently occurring outcomes

- Identification of duplicate transactions

- Identification of missing values

- Validation of data

The general industry practice is based on investigating leads based on fraud patterns using relevant information. Let's assume we want to investigate the billings of optometrists in Oregon. One of the first things to do can involve choosing a subset of optometrists that bill similarly for patients that have similar health risk profiles.

Then, you can compute the average payment of each doctor per beneficiary and check how different each doctor is from the overall average. This is a simple example of a descriptive statistical analysis method. However, working with the averages may not be enough. We are also interested in the variability.

In order to measure dispersion among doctors, statistics such as standard deviation and variance are widely used. These capture the cumulative differences of each observation and the mean. The higher the standard deviation is, the more spread the data are. How far should we go away from the mean to suggest that the billing behavior is suspicious? Let's assume the mean spending by optometrists and pediatricians for similar patients is $100 per beneficiary. One optometrist spent $110 per beneficiary whereas the spending of a particular pediatrician is $120. How do we compare these two providers who are different from their peers? If we have resources to investigate one of them, which one would we choose?

Do we simply choose the one with the higher spending? Not necessarily. The variations of spending by each provider type is also important.

One option is to use a standardization to rescale each variable. This is based on subtracting the mean from the outcome of interest and divide it by its standard deviation. This is also referred to as the Z-score. It indicates how many standard deviations the observation of interest is from the mean. The mean value of the data set always has a Z-score of 0. The so-called empirical rule is proposed for the averages of the symmetric distributions, showing approximately 68%, 95%, and 99.7% of the values lying within one, two, and three standard deviations of the mean. Therefore, observations that have a Z-score value that is larger than 3 or smaller than -3 are very unlikely and can be defined as so-called outliers. Z-scores can also help in comparing observations from different data sets that may be investigated with respect to different variables. More on outlier detection methods is presented in Chapter 5.

Back to our original question; if we use standardization, the standard deviations also become important. Let's assume that the standard deviations of spending per beneficiary by optometrists and pediatricians are $5 and $40 respectively. We can argue that the billing behavior of the pediatrician is more within the acceptable norms, and similar to other pediatricians since it has a smaller Z value. Therefore, the billing behavior of the optometrist can be deemed as more suspicious.

Another option is to compare the measures of probability distributions of each variable. A probability distribution lists each outcome of a statistical experiment or random variable along with the frequencies of occurrences. For instance, the left panel of Figure 2.1 presents information about the distribution

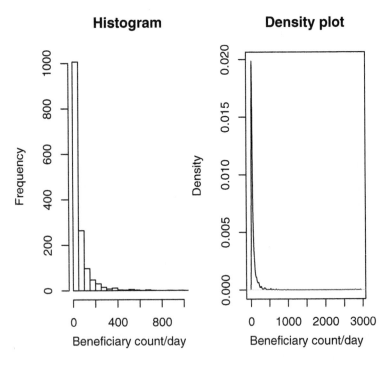

FIGURE 2.1   Histogram and density plot of beneficiary count per day.

of the number of distinct Medicare beneficiaries per day served by a particular provider. This figure, a so-called histogram, simply presents a chart of counts in chosen ranges. Histograms are used to compare the relative frequencies of data points. The height of each bin corresponds to the frequency of the relevant range of outcomes.

Whereas the right panel of Figure 2.1 presents a density plot which presents the actual probability densities of observations. It is like a continuous histogram where the area under plot is equal to 1. When you look at a density plot, you are more interested in the overall shape of the curve than in the actual values on the y-axis, which represents the density.

Boxplots are another set of visualization tools that display certain summary statistics and reveal the distribution of a variable. Furthermore, they can be used as indicators of the relationships among variables. Before discussing them further, first, let's introduce percentiles. They provide a flexible way of comparing the rank of an observation within a data set. For instance, the 25th percentile refers to the value that is greater than 25% of the observations, and also is referred to as the first quartile (Q1). Similarly, the 75th percentile or the third quartile (Q3) is the value that is greater than 75% of the observations.

Boxplots generally report the lower and upper whisker values, median, Q1, and Q3. Figure 2.2 presents the boxplot of the average Medicare payment amount per claim for optometrists in Oregon. The width of the box is computed as the difference of Q3 and Q1, also known as the inter-quartile range (IQR). The whisker values provide an understanding of the range of values within the norms. There are a number of different methods proposed to compute these values—we will not get into those details.

Observations that are outside the range of whiskers are denoted with circles and can be referred to as outliers. In order to compare a numerical variable with respect to a categorical variable,

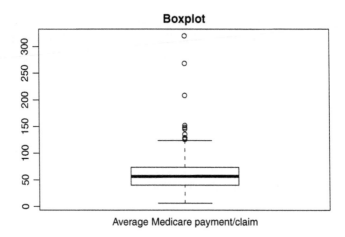

FIGURE 2.2   Boxplot of average payment per claim for optometrists.

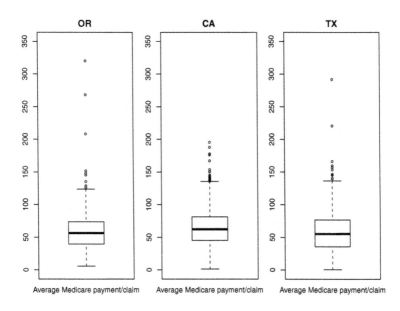

FIGURE 2.3   Conditional boxplot of average payment per claim for optometrists of different states.

conditional boxplots can be used. For instance, Figure 2.3 shows the boxplot of average payments for optometrists that are billing in Oregon (OR), California (CA), and Texas (TX). The distributions with respect to states can be compared to detect the outliers. Looking at this conditional boxplot, the variability in Texas seems to be slightly higher, while a few of the outliers in Oregon and Texas are dramatically different from the rest.

A very crucial aspect of evaluating figures is scale. For instance, the scales in boxplots are preferred to be the same for straightforward visual comparisons. This is a trick that is used to manipulate the power of visuals. For instance, side by side boxplots with different scales can easily be deceptive.

Lastly, scatterplots can be used to compare two numerical variables. For instance, the average Medicare payment amount per claim of each provider and the number of distinct Medicare beneficiaries per day served by that provider can be compared using the scatterplot in Figure 2.4.

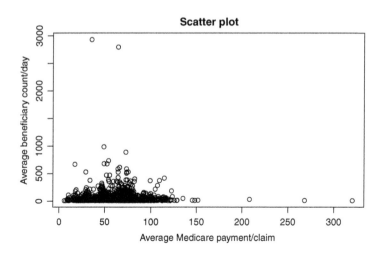

FIGURE 2.4 Scatterplot of average Medicare payment amount per claim and number of beneficiaries served per day.

DISCUSSION

Data pre-processing and security are and will be integral parts of any type of health care data analysis. Descriptive statistical analysis is widely used as the first step. Those methods are especially beneficial and widely used where large amounts of overpayment can be easily identified due to irrefutable evidence. Such examples include providers who bill for patients that are logically impossible to serve, for instance, dead beneficiaries. Providers who are outliers in terms of their services can easily be identified. Widely used industry tools for descriptive statistical analysis include spreadsheet-based software. Commercial tools that provide nice visualizations and let the user perform such computations within the database are also becoming popular. Open source statistical software such as R can also be utilized for many statistical tasks.

Clinical insights are very valuable while evaluating the results of the descriptive statistical tools. Although the disease and respective treatment intensity in each area and for each provider type are expected to be similar, potential differences should be analyzed with caution. Understanding patient characteristics is important and generally measured by variables such as risk-adjusted patient profiles that consider the demographical information of the patients and geographical realities of a particular location. Descriptive statistics can be employed separately for each defined category of variables such as region, patient profile, or provider type.

We should note that the aforementioned tools are among many other widely used descriptive statistical methods. Our discussion is by no means exhaustive. For instance, comparison of the billing behavior of providers over years is a widely used method. A drastic increase or drop in a year can be a sign of fraudulent behavior and may be flagged for further investigation. Such trend analysis is often utilized in industry.

Descriptive methods are also effective to generate leads within rule-based algorithms especially for some well-established cases.

However, sophisticated fraudsters can find ways to circumvent them. In addition, descriptive method-based rules may not discover new patterns of fraud and abuse automatically and are difficult to maintain and update over time. The variable structure of data and the nature of fraud motivate the widespread use of data mining methods. The following chapters present overviews of sampling and data mining methods that can address these concerns.

KEY TAKEAWAYS

1. Health care data are generally classified into practitioners, clinical instances and claims data.

2. Data sources such as patient-generated data, public records, web queries, social media, and weather events can be used complementarily with health care data.

3. There are a number of publicly available health care data sources that are provided by the CMS and the OIG.

4. Ensuring data quality is crucial before the application of any statistical method, and a number of data pre-processing and transformation methods are proposed to help with that.

5. Health care data security is protected by a number of federal acts such as HIPAA.

6. Descriptive methods are used as the first step of many statistical approaches and generally include measures of location, dispersion, and distribution.

7. Outliers are observations that do not represent the general characteristics of the population and are far away from the mean.

8. Descriptive statistics–based methods are especially beneficial within rule-based algorithms.

9. Visualization techniques such as histograms, boxplots, and scatterplots help compare and analyze the data sets.

ADDITIONAL RESOURCES

1. Health care claims data resources, download, and interactive visualization:

   CMS (2018). CMS research statistics data and systems. The Centers for Medicare & Medicaid Services. https://go.cms.gov/2J9oSVk

2. R reference and free download:

   R Core Team (2016). R: A Language and Environment for Statistical Computing. R Foundation for Statistical Computing, Vienna, Austria. https://www.r-project.org/

CHAPTER 3

Sampling and Overpayment Estimation

OVERVIEW

When government audits reveal that Medicare has overpaid health care claims to private insurance companies, it's big news. And for good reason: overpayment reflects fraud, inefficiency, or human errors, and it wastes the money of taxpayers. The audits themselves can be costly and time-consuming.

In January 2017, for example, the nonprofit, nonpartisan Center for Public Integrity reported on government audits showing that Medicare had overpaid five insurance plans a total of $128 million in just one year, in 2007. In September 2017, the Department of Health and Human Services reported that between 2013 and 2016, Medicare had inappropriately paid acute-care hospitals more than $50 million—for outpatient services that weren't even done at the hospitals. In 2015, analysis of 25 million Medicare claims resulted in the finding of erroneously estimated hospital costs. The agency overpaid hospitals $2.6 billion between 2009 and 2013.

Health care overpayment investigations have big stakes, not only for insurance companies but also for patients, taxpayers, and policymakers. However, accurately measuring the problem and putting a value on overpayments brings its own complicated procedure. Insurance companies and Medicare handle so many insurance claims that it's impractical and cost-prohibitive for auditors to analyze every claim, searching for signs of financial mismanagement.

So what's the best way to solve the problem in a cost-effective way? Use statistics. Audits can be expensive, so auditing agencies have adopted ways to analyze subsets of big data sets to make inferences for the entire group. Those subsets are called samples, and the results of sample investigations are projected, or extrapolated, to the entire population. This process is called sampling.

In this chapter, we will use the timely example of Medicare audits as an illustration of how analysts conduct population sampling, analyze the sample, and extend their findings to the entire population. Such probability sampling in health care investigations has been accepted as part of the legal framework in the U.S. since 1986.

The U.S. government requires auditors to follow certain guidelines to ensure the accuracy of the resulting overpayment estimates. If the insurance companies claim to find faults in the analysis, they'll push back against being asked to repay—which is exactly what happened with the $128 million overpayment, as we'll see at the end of the chapter.

Such an audit is designed to arrive at a fair amount of recovery, but also to identify patterns of overpayment in the system— certain claims or companies that are more likely to be involved, for example. The ultimate goal, of course, is to inhibit the culture of sustained overpayment and get billing right the first time. Managing the trade-off between audit costs and overpayment accuracy is one of the main challenges in the effort, and the details of the sampling procedure are very important because the investigator needs to guarantee their validity in potential court hearings.

This chapter provides an overview of the use of probability sampling methods and subsequent overpayment estimation methods in health care fraud assessment. We'll address the following questions:

- Why do we sample?

- What are the main sampling types used in health care fraud assessment?

- How do we choose the sampling method for a particular medical audit?

- After retrieving a sample, how do we use that information to make an inference?

- What are the common tools that are used to conduct sampling and overpayment estimation?

SAMPLING AND OVERPAYMENT ESTIMATION

Between 2005 and 2008, the U.S. government conducted a test run of a new national program designed to identify cases where improper health care services had been made to providers. That demonstration was limited to only six states, and it led to the identification and recovery of $900 million in overpayments, which were returned to Medicare. The program works both ways, and during the same time period audits identified $38 million in underpayments, which were returned to health care providers.

In January 2010, the program was expanded to the entire country, and since then the Centers for Medicare & Medicaid Services (CMS) has conducted regular audits of providers, including hospitals, physician practices, nursing homes, home health agencies, medical equipment supplies, and others. According to the CMS, errors can be revealed by using a variety of sources, including probe samples, data analysis, provider history, information from law enforcement investigations, allegations of wrongdoing by employees of a provider, and previous investigations by the Office of Inspector General (OIG) within the Department of Health and Human Services.

SAMPLING PROCEDURES

Many research areas use sampling and extrapolation to find out about a population without examining all the data points. For example, to estimate the spread rate of the flu, a researcher might survey a subset of the population and then scale those findings up. Prior to an election, political analysts use polls that are retrieved from a subset of the population to understand public behavior. Marketing professionals use focus groups or market studies and collect data on a small number of people of interest. Then, these results are used to understand the behavior of the target population.

Sampling can be used in health care fraud investigations only after the demonstration of a sustained or high level of payment error, according to the Medicare Prescription Drug, Improvement, and Modernization Act of 2003. In other words, you need to have enough justification for the sampling audit.

Sampling is conducted in health care fraud assessment because it would be too consuming of time and resources to evaluate each claim filed by a provider. A typical investigation may involve tens of thousands, or even hundreds of thousands of claims, or more. To assess the legitimacy of an individual claim, an auditor would need to spend extensive time going through medical records and other documents. In some cases, they may interview the patients and providers involved in the case.

By using sampling, investigators can get a representative measure of the overpayment amount for a subset, and then use that to estimate the overpayment amount for the whole population. But obtaining a sample isn't as easy as picking a few claims and multiplying.

The investigation process follows these steps:

1. Selection of the provider to be evaluated.

2. Selection of the investigation period. During which years should claims be analyzed?

3. Definition of the population to be investigated. Statisticians refer to this population as the "universe" under investigation.

4. Determination of the sample unit. How many claims—or parts of claims—should be analyzed to provide an accurate representation of the universe?

5. Determination of the sampling frame, which is the list of items to be evaluated. In Medicare investigations, the sampling frame may include individual lines within claims.

6. Design of the sampling plan. There are many designs used for sampling, which will be discussed in this chapter.

7. Selection of the sample.

8. Review each case within the frame and compute the overpayment amount.

9. Add up the overpayment for each claim to find the overpayment for the sampling frame.

10. Use the overpayment estimate for the sample to compute the overpayment for the universe.

Each of these steps influences the outcome, but the sixth one—choosing an appropriate sampling design—can have a tremendous effect not only on the overpayment estimate but also on the trustworthiness of the study. Not every sample design is appropriate for every situation, so auditors have to know the strengths and weakness of each. Next, we present widely used sampling designs.

Simple random sampling has been widely used in health care audits. It's fairly intuitive to grasp, and its outcomes are straightforward to communicate to others. It's exactly what it sounds like: every unit—or claim, in this case—has the same probability of being chosen. The auditor chooses a set number of units, randomly, and no unit can be chosen more than once. This is called sampling without replacement. Simple random sampling eliminates the risk of having a biased sample. It generally results in high precision when the population is homogeneous, which means the sampling units are similar to each other. For instance, an auditor might choose to look at claims from the same procedure. This method may be preferable when the investigator lacks initial information about population, such as potential subgroups or patterns in claims.

Systematic sampling is based on selecting a random starting unit and then choosing the rest of the sample using a systematic rule. For example, an auditor might choose every fourth or fifth claim, or claims separated by a certain number of days. This design ensures that a sample includes units that are distributed evenly through a population. It's straightforward to implement. However, there is the potential for unpredicted biases, since only certain types of claims may be selected which may compromise the randomness.

If the population is heterogeneous, which means the sampling units are dissimilar from each other, then random and systemic sampling may be imprecise, and favor some subsets over others. For these populations, an auditor may choose to use stratified sampling. This design begins with separating a population into mutually exclusive, homogeneous groups using a stratification variable. Stratification variables used in Medicare audits include payment amounts and medical procedure codes. The objective is to form groups in a way that minimizes the differences within each group and maximizes the differences from group to group.

Separate estimates of the overpayment are made for each group and then weighted according to how the groups were formed. Combined, they give an overpayment estimate for the whole population.

Its overall success depends on the accuracy of the group formation. However, if done right, this can provide a smaller margin of error as well as additional information about each group. In addition, with stratified sampling, an auditor can use different sampling designs within each group for cost or other considerations.

Overall, stratification is often the method of choice if an auditor has access to sufficient information to break up a population into homogeneous subgroups.

Cluster sampling is similar to stratified sampling. It involves choosing groups, or clusters, with respect to similarity, easiness, or cost. Then, the auditor can choose a simple random group of clusters to evaluate. In a similar context, a particular county can be chosen for election polling. For example, a county of 2,000 people in a swing state can be used to make estimates about more than 200 million voters.

Similarly, an auditor may separate all the claims from a region into clusters of hospitals, and then evaluate a random sample of a given cluster. That cluster of hospitals is assumed to reflect the characteristics of the region. That is why cluster sampling is usually inexpensive to conduct.

Table 3.1 summarizes the list of situations and potentially preferred sampling methods for those cases. Auditors choose the appropriate sampling design based on the number of claims and how much information is accessible within those claims, and from the provider under investigation.

TABLE 3.1   List of Cases and Proper Sampling Method

Situation	Sampling Method
Lack of advanced auxiliary information of the population	Simple Random
Concern of compatibility with a given statistical software	Simple Random
Need for having the ability to compare subgroups	Stratified
Concern of retrieving smaller margin of error for the same sample size	Stratified
Need for utilizing knowledge of the population subgroups	Stratified
Need for different statistical procedures for each subgroup	Stratified
Concern of correlated (linearly related) samples	Systematic
Evidence of pre-existing classification of providers	Cluster

A CLOSER LOOK AT STRATIFIED SAMPLING

Here's an example of how auditors choose a sampling method. Imagine a team of auditors charged with evaluating tens of thousands of outpatient claims from a hospital under evaluation, and they have to figure out how to sample the population in the best way. Before they decide on a method, they have to get some insight into the population, so they first execute a quick graph of how much money was paid to individual claims.

Figure 3.1 shows many peaks, which statisticians describe as being "multi-modal." It shows that patient claims more or less fall into natural groups, represented by those peaks. In this case, if the auditor thinks the overpayment percentage is relatively constant across different payment levels, she may want to take advantage of the emerging natural layers. This is especially relevant if the auditor has a particular interest in claims at different payment levels.

The auditor would use the payment values to organize the data into different layers. The strata, or layers, could be determined by

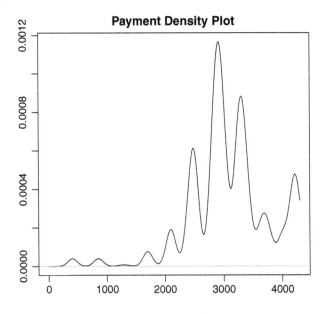

FIGURE 3.1   Density plot of payment population.

payment data, where each subsequent stratum represents a higher payment amount. The CMS has approved auditors to conduct stratified sampling with five or six strata, although other sizes do not make the sampling process invalid.

At this point, the auditor has to make concrete decisions about the strata. They have to determine the boundaries, which will determine which claims go into which group, and the sample size for each stratum. For example, an auditor may choose proportional allocation, which means the size of the sample drawn from each stratum is proportional to the relative size of the stratum in the population. Imagine a case where the claims are divided into three groups, where the first group has 10,000 claims, the second group has 20,000 claims, and the third group has 30,000 claims. If the auditors wanted a sample of 60 claims, they might use proportional allocation to select 10 claims from the first group, 20 from the second, and 30 from the third.

Proportional allocation can be precise if the groups are big enough; if they are too small, they may skew the overall sample.

In those cases, the auditors may need to use one of the disproportional stratified sampling methods. If the aim is to focus on a particular stratum, the investigator may want to ensure that the sample size drawn from that stratum is large enough. If the objective is to conduct inter-strata analysis, then an equal number of samples can be chosen for each stratum.

As an alternative, the optimum allocation method allows the auditor to consider both the precision of the estimates and the sampling cost. In this approach, the auditor samples more heavily from strata with high variance and large population size, in which the cost of sampling is low. Even if costs are not available, or they're equal from all strata, optimum allocation can be changed to only consider some other variable. Statisticians call this practice the Neyman Allocation, but that's pretty far in the weeds, and we won't discuss it here.

Auditors are not confined to just one of these designs; depending on the data set and the goal of the investigation, they may need to combine many. An auditor may initially identify a sample of units proportional to their size—focusing on a single hospital, for example—and then divide that sample by provider type.

After partitioning a large population into homogeneous subsets, an auditor can reach a point where they can draw random samples to comprise the sample set to be evaluated. This kind of sampling may seem laborious, but it can help an auditor find a balance between precision and cost.

Some statisticians have suggested modifications to this multistage approach that might save time and money. Starting with a small random sample to see if the number of overpaid claims exceed a given threshold may be a good starting point. If the number of overpaid claims is low, then the health care investigation might be dropped—especially if the cost of the investigation exceeds the cost of the overpayments that might be recovered.

Other statisticians have suggested multistage sampling frameworks that are tuned using other tools. For example, tools from information theory—an interdisciplinary field that focuses on how we understand, store, and transmit information—can be used. An auditor may first evaluate the information content of prospective samples within a stratified sampling framework. Then, using that information, they can pick one representative stratum to learn about the overpayment.

So far, we have talked a lot about the methods and the reasoning behind them. In practice, it goes something like this: an auditor uses one of the many statistical software options. Here, we use an open source statistical computing program called R—that anyone can use—to put these methods into practice.

This is a bit in the weeds, but it should help illustrate the application. Imagine you have a universe of 8,278 claims, and you want to divide those claims into strata based on payment amount. You can use one of the discussed methods and determine the boundary values for your payment strata. The descriptive statistics of each group may look like Table 3.2.

Each line of the table represents one stratum, and each stratum represents a payment bracket higher than the one before it. The

TABLE 3.2 Descriptive Statistics of Payment

Stratum	Mean	Std. Dev.	2.5%	25%	Median	75%	97.5%	N_s
s = 1	69.40	46.92	20.00	30.00	40.00	40.00	50.00	3949
s = 2	915.69	180.87	50.00	50.00	70.00	90.00	100.00	1402
s = 3	2335.37	244.61	1700.00	2100.00	2500.00	2500.00	2500.00	588
s = 4	3076.60	206.76	2800.00	2900.00	3000.00	3300.00	3400.00	1675
s = 5	4012.65	246.71	3600.00	3700.00	4100.00	4200.00	4300.00	664
Overall	1298.47	1441.22	30.00	50.00	600.00	2900.00	4200.00	N = 8278

first column reports the arithmetic mean, or average, payment in that stratum. The fifth column reports the median, which is the middle value of the set. The mean and median payments are quite different from group to group, which gives the auditor confidence that these groups really are distinct from each other. A variety of percentiles, as well as the standard deviation of payment values in each stratum, are also presented. These would give an idea about the payment distribution.

To pick samples from each group, the auditor would use another piece of statistical software called Rat-Stats, which is publicly available and provided by the Office of the Inspector General. It performs three main functions. First, it can determine the sample size for a given level of accuracy. Second, it uses a certified random number generator—yes, it has to be certified in order to hold up under legal scrutiny—to pick the claims to be evaluated. And third, it performs statistical calculations on the selected sample, so that they can be generalized for the population of interest.

Finding the right sample size requires that an auditor balances resources, such as time and money, against the rigor and precision of the results. She wants to feel confident that her findings are valid; at the same time, she has to do it on a budget.

Rat-Stats can conduct important operations on the data, such as attribute appraisal and variable appraisal. Attribute appraisals help estimate how often a given condition occurs—for example, the percentage of overpaid claims from a particular hospital. Variable appraisals measure other characteristics of the sample, like estimating the average amount of overpaid claims and extrapolating that to the whole population. These are critical in health care overpayment investigations.

Rat-Stats was built and improved by statisticians, and it can accommodate and even combine the many different designs we've already discussed, like cluster and stratified sampling methods. It allows users to consider cases where they know the sample size, and where they don't. The program also reports confidence levels for different statistical approaches, which can help auditors find the most precise approach. In this case, precision refers to the margin of error in a calculation. The margin of error is usually reported as a percentage higher or lower than the mean.

This last feature is important. For Medicare audits, the Office of the Inspector General may recommend at least a 90% confidence, and a precision no worse than 25%, for the analysis. These recommendations determine how many claims an auditor will have to investigate.

As a test case, imagine we have the same data we looked at in Table 3.2, and our fictional auditors have to identify an appropriate sample size. Again, they are looking at a collection of 8,278 claims. When they use the software to identify the strata and run the analysis, both on individual groups and on the data set in its entirety, Figures 3.2 and 3.3 are examples of what the program would return. These tables are less intimidating than they look, I promise.

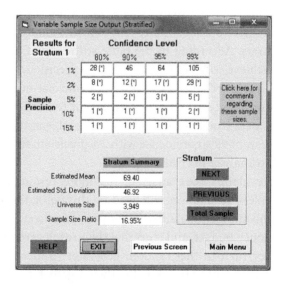

FIGURE 3.2 Rat-Stats output screen with sample precision and confidence levels for a given stratum.

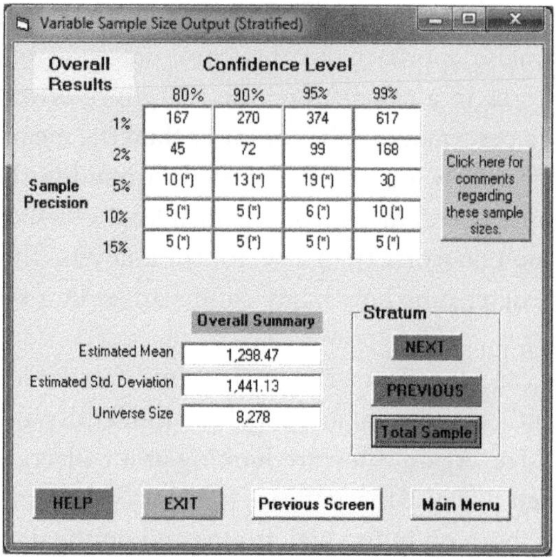

FIGURE 3.3   Rat-Stats output screen with sample precision and confidence levels for all strata combined.

The box on Figure 3.2 analyzes data from a single stratum. You can imagine there are four more, very similar looking boxes, for the other strata. According to the table at the top, if an auditor wanted to achieve 99% confidence with 1% precision, she would need to analyze 105 claims out of the entire universe of 3,949.

The box on Figure 3.3 reports statistics for the entire data set. It combines data from all the strata, and the table at the top tells us that if we want 99% confidence with 2% precision, for example, we'd want to analyze 168 claims, total.

The more claims they analyze, the more resources—time, money, people—the auditors will need. At this point, they have to decide on the number of claims to study in order to stay within a time and money budget, and trust that their results are valid.

Then, they turn to that random generator to identify which claims they'll be analyzing. Finally, they'll gather those claims and go through them, line by line, to see if the amount paid by Medicare to the hospital was appropriate.

OVERPAYMENT ESTIMATION

We've already seen that auditors can choose from among many designs to sample a population. They can also use a variety of tools to estimate the overpayment—and feel confident that the estimate is a good one as long as they follow the rules of statistics. The easiest way would be to simply calculate the mean overpayment for a sample, and then multiply that mean overpayment by the population size. If the mean overpayment estimate is $10 for a sample representing 1% of the population, then the full overpayment estimate would be $1,000. That's called simple expansion. Another approach is the ratio estimator. In this case, the auditors multiply the total payment value—for all claims in the population—by the ratio of the sample overpayment to the sample payment value. For example: Say auditors find an overpayment of $50 in a sample where the payment value was $1,000. If the total payment for the whole population was $10,000, then the total overpayment estimate would be $10,000 \times (50/1000) = 500.

Having introduced these, recovering that money isn't as simple as sending a bill to the provider for the estimated average amount. To protect the providers in case of erroneous estimation and to give them the benefit of the doubt, the government requires auditors to subtract the margin of error from the mean amount. This practice may mean that Medicare doesn't recover all of its funds, but that the audited provider is protected with a high level of confidence.

There are underlying assumptions behind using margin of error, though. You assume that either your sample size is large enough to be representative of the whole population, or that the whole population follows Normal distribution. Normal distribution is based on the idea that data are evenly spread around the mean (and the middle [median] value). This is how statisticians quantify the idea that most claims have payment values near the mean, or average. Only a small fraction of the claims would have payment values that differed quite a bit from the mean.

However, health care data are notoriously skewed and do not reliably follow Normal distribution. So, the auditors have to get sample sizes big enough to make a good estimate because small sample sizes won't reflect the whole picture. You can see the tension here because it is the same as at every other step of the process: Auditors have to find a balance between rigor and cost. Bigger sample sizes bring higher costs, which decrease the net gain from the overpayment recovery.

There are many ongoing efforts to develop alternative models that can deal with these types of health care claims. Some of the newer ideas include using non-Normal distributions to estimate the overpayment and explicit modeling of overpayments in different groups. We will not get into details in this book. Most of these methods haven't yet been validated for use in Medicare audits, which means they are still in the experimental stage.

Which leads us back to Rat-Stats, which auditors use to estimate the overpayment for each stratum, and then scale up that amount to the entire population. Because we know that our sample size gives us the confidence and precision required for the estimate, we can use the statistical methods.

At last, the auditors have a recovery number that accurately estimates the overpayment amount; it's that number that they take to the prosecutors, who will approach the providers.

DISCUSSION

Probability sampling and overpayment estimation are well-established methodologies that can be used in health care fraud assessment frameworks. However, in the audit is just the beginning of the story. It's often followed by months, if not years, of legal, financial, and policy wrangling.

Courts may put a hold on the recoveries that are computed by extrapolation if the accuracy of the audits and methods is challenged. Sometimes, government agencies may not be confident going to court to defend their findings. In the case I mentioned at the beginning of this chapter, the CMS used extrapolations to estimate that Medicare overpaid five insurance plans by $128 million. The plans were reported to be overstating the severity of medical conditions such as diabetes and depression; in some cases, they had filed claims for treatment of diseases that were never diagnosed. However, the insurance industry pushed back. Eventually, the CMS settled for a few cents on the dollar, recovering just under $3.4 million.

Providers can appeal extrapolation and recovery decisions. Providers try to avoid repaying the cost by arguing that the extrapolations include low sample size and that there was a lack of initial evidence of wrongdoing. The concerns about the low sample size are generally dismissed—since CMS guidelines and court rulings confirm the results with statistically valid probability samples.

Providers can appeal the payments through federal district courts, the Medicare Appeals Council, and the relevant state courts. However, according to the American Hospital Association (AHA), such appeals can take three to five years to pursue given the two-year moratorium on assigning new claim appeals to administrative law judges.

Other countries use similar approaches to try to recover overpayments made by government agencies. In Australia, for example, the Professional Services Review (PSR) was established in 1994 to apply rigorous standards to audit services with the main goal of protecting the integrity of the programs. To address the practical difficulty of examining all services, statistical sampling was incorporated into PSR schemes. Each assessment has to include at least 25 claims, and the audit has to achieve at most a 10% margin of error, with 95% confidence. As in the United States, the estimated percentage of overpayment is decreased by the margin of error to retrieve the final recoupment amount.

Recent cases in the United States—where Medicare fails to recover any significant fraction of an overpayment amount—show that the system often breaks down, and taxpayers pay the price. Policymakers, providers, and courts need to be more familiar with the statistical validity and the strength of these models, and find ways to fix the flaws in the Medicare system.

KEY TAKEAWAYS

1. Managing the trade-off between audit costs and accuracy of the overpayment estimation is one of the main challenges in sampling resource-allocation decisions.

2. Sampling and overpayment estimation help health care auditors analyze large data sets of health care claims.

3. The appropriate sampling design depends on the available information of the population and the goal of the investigation.

4. The recoupment amount in the U.S. is computed by subtracting a margin of error from the average overpayment estimate, and generally computed so that the provider is protected for erroneous estimates with a certain level of confidence.

5. To increase the use of sampling and overpayment estimation for extrapolation, the providers and the courts should be educated more about embracing uncertainty and statistical validity.

ADDITIONAL RESOURCES

1. A classical sampling textbook

    Cochran, W. G. (2007). *Sampling techniques.* John Wiley & Sons. New York, NY.

2. Current sampling guidelines in the U.S.

    CMS (2011). Medicare program integrity manual chapter 8 administrative actions and statistical sampling for overpayment estimates. The Centers for Medicare & Medicaid Services. https://go.cms.gov/2LXmQFx.

3. Rat-stats free download and user guide:

   OIG (2016). Ratstats user guide. Office of Inspector General. http://oig.hhs.gov/organization/oas/ratstats/UserG uide2010_04js.pdf.

4. A website with comprehensive sampling resources:

   Yancey, W. (2012). Sampling for Medicare and other claims. http://www.willyancey.com/sampling-claims.html.

5. Novel approaches for sampling and overpayment estimation which may require more advanced statistical understanding:

   - Bell, R. and Nicholls, D. (2006). Statistical sampling in a legislative framework for peer review of medical services. *Journal of the Law and Medicine*, 14(2):209–19.

   - Edwards, D., Ward-Besser, G., Lasecki, J., Parker, B., Wieduwilt, K., Wu, F. and Moorhead, P. (2003). The minimum sum method: a distribution-free sampling procedure for Medicare fraud investigations. *Health Services and Outcomes Research Methodology*, 4(4):241–63.

   - Ekin, T., Musal, R. M. and Fulton, L. V. (2015). Overpayment models for medical audits: multiple scenarios. *Journal of Applied Statistics*, 42(11):2391–405.

   - Ekin, T., Ieva, F., Ruggeri, F., Soyer, R. (2018) Statistical medical fraud assessment: Exposition to an emerging field. *International Statistical Review*. 86(3), 379–402.

   - Guthrie, D., Birnbaum, Z.W., Dixon, W.J., Feinberg, S.E, Gentleman, J.F., Landwehr, J.M. . . . Williams, S.J. (1989). Statistical models and analysis in auditing: Panel on nonstandard mixtures of distributions. *Statistical Science*, 4:2–33.

   - Ignatova, I. and Edwards, D. (2008). Probe samples and the minimum sum method for Medicare fraud investigations. *Health Services and Outcomes Research Methodology*, 8(4):209–21.

- Musal, R. and Ekin, T. (2018). Information theoretic multi-stage sampling framework for medical audits. *Applied Stochastic Models in Business and Industry*, 34(6):893–907.

- Musal, R. and Ekin, T. (2017). Medical overpayment estimation: A Bayesian approach. *Statistical Modelling*, 17(3):196–222.

CHAPTER 4

Predicting Health Care Fraud

OVERVIEW

Data generation, storage, and analysis have become more afford-able especially in the last decade. Data are being produced at large quantities and high speeds thanks to devices such as sensors or channels such as social media. These factors coupled with the increase in the power and affordability of computing resources, make statistical and data mining methods an integral part of the business processes in many industries. Different buzzwords are used to describe this emergence of analytical methods. One of them, so-called business analytics, can be defined as involving the use of quantitative methods to reveal information from data and create value for businesses. When the data are too large or complex that they cannot be handled with traditional software, it is also referred to as "big data analytics."

Actually, the relevant analytical methods of statistics, data mining, and machine learning have been around for quite some time, but their popularity and business applications have increased tremendously. Nowadays, we do our shopping online with the help of recommendation algorithms. We let our online streaming provider choose a movie for us. We listen to the song that is picked by our personalized radio station. The mainstream use of self-driving or assisted cars is closer to becoming an everyday reality. These developments result in a need for new data mining and analytics software and new methods. That is why analytics has become a scholarly and commercial area of interest. Health care fraud assessment is no exception. Current health care databases have a huge number of records. Relatively simple descriptive data analysis tools may not be capable of handling such big data. Therefore, health care fraud analytics is becoming more commonplace worldwide.

Put yourself in the shoes of an auditor who oversees health care claims submitted in Texas Medicare. There are more than three million Medicare enrollees who are provided health care service in more than 6,000 hospitals. These service provider facilities do not even include skilled nursing facilities, hospices, or independent and clinical labs. This auditor's responsibilities do not even include Medicaid or any other governmental insurance programs beneficiaries, nor the remaining 49 states. Nevertheless, she has to make decisions on how to choose the providers, beneficiaries, or claims to audit. If it were you, how would you use your time to discover instances of health care fraud? In this chapter, we will start exploring possible answers to that question by presenting predictive methods.

One option to initiate investigations is using the information and leads generated by telephone tip lines. Indeed, these tips are very beneficial for many cases. But is this not very passive, even based on luck and good wishes, especially since the auditor could be dealing with organized crime? Should the auditor wait until someone picks up the phone and blows the whistle? Moreover, such leads for further investigations are mostly generated after the payment is processed. Even if any wrongdoing is found, the mechanism to recover the overpayment is initiated after the fact of fraud. That is why these so-called pay-and-chase methods may not be very helpful for more organized criminal activities. Professional criminals can hide their identities and relationships, and disappear after a few successful transactions.

Have you ever heard about Angel Lagoa? He is among the most wanted fugitives in the United States. Mr. Lagoa owned Trinity Senior Care, Inc., located in Miami, Florida. According to the Office of Inspector General investigation, Lagoa and his co-conspirators recruited patients for their beneficiary information in exchange for illegal kickback payments. Then, they submitted approximately $30 million worth false or fraudulent claims to Medicare. He was involved with fraud between 2010 and 2015. However, his activities went unnoticed for too long, and he fled just before getting caught.

How could we have understood what was really going on, and reacted before losing $30 million? As discussed in Chapter 2, descriptive statistical analysis could proactively describe the patterns in health care claims data. However, they may not be as successful at revealing patterns within large and complex health care claims data sets. Predictions of the fraud probability of a claim or of a particular provider could be beneficial to rank claims and determine which ones to audit. Therefore, these predictive methods can be effectively used during both phases of pre-payment and post-payment.

Regression and classification methods are frequently used as predictive methods. Each algorithm has its own strengths and weaknesses, and can be preferred for certain cases. Although they are successful in revealing many instances of fraud, ensuring their accuracy is critical.

Prediction of legitimate claims as fraudulent would waste time and resources since innocent providers have to go through an investigation. None of the parties involved would be content with such an outcome. Fraudulent claims going unnoticed may mean fraudsters getting their way and tax money being spent on over-payments. Such trade-offs related to accuracy are important in health care audits.

Some of the questions to be addressed in this chapter include

- How are analytical methods used in the grand scheme of health care fraud assessment frameworks?

- What is a predictive method, and how can it be used for health care fraud detection?

- What is regression? How can it be used for prediction of overpayment amount and fraud probability?

- How can we utilize classification algorithms with health care claims data?

- How we can measure the accuracy of predictive methods?

HEALTH CARE FRAUD ANALYTICS

Analytical methods are mainly used for both pre- and post-payment fraud audits. Pre-payment processes are designed to deny clearly fraudulent claims and suspend suspicious claims for investigation, while considering the standard variability of medical care. These systems are generally conducted using a set of rules and thresholds. For instance, if a patient visits a particular hospital ten times in a given month or if the patient receives frequent services from a facility more than 160 kilometers (approximately 100 miles) from their residential address, simple rules can flag such billings.

Pre-payment reviews are crucial since they are the first line of defense for fraud prevention systems. In general, recovering lost revenue is labor intensive and can be an unsuccessful venture. Efficient use of pre-payment analytics results in recovering all potential payment while removing the need for investigations. Pre-payment systems can also provide further input and feedback for the post-payment reviews.

The limited resources and growing amount of sophisticated fraud schemes prevent pre-payment reviews from capturing all potentially fraudulent claims. The auditors generally have up to 14 to 30 days to postpone the payments before conducting additional review. In addition, identification of new patterns of irregular behavior is a slow process and requires input from many auditors and reports. As a result, a significant amount of fraudulent health care claims can pass the pre-payment review and get paid out before they can be flagged.

As the secondary line of defense, post-payment reviews are used to detect fraudulent activities and prioritize leads for future audits. Ideally, pre-payment and post-payment systems would provide feedback to each other in an integrated manner and help the policymakers modify the regulations and rules accordingly.

The developments in the industry and growing data sources prompted the U.S. federal government to explore the application of statistical and analytical methods for health care fraud detection. The Congress appropriated $100 million to the Centers of Medicare & Medicaid Services (CMS) via the 2010 Small Business Jobs Act. CMS has introduced their analytical system, the so-called Fraud Prevention System (FPS), for health care fraud assessment in Medicare.

Tools such as the Fraud Prevention System help analyze massive amounts of data and provide useful and interesting information about patterns and relationships that exist within the data that might otherwise be missed. Such comprehensive application of analytical algorithms helps the CMS to investigate signals of fraud before and after the payment is made. This can prevent improper payments in addition to using the traditional "pay-and-chase" methods. Particularly, provider networks, billing patterns, and beneficiary utilization patterns are analyzed in order to detect patterns that represent high risks of fraudulent activity.

Figure 4.1, which is taken from the *CMS 2014 FPS Implementation Report* (CMS, 2014), summarizes the overall fraud assessment procedure. The deployment of the model on health care claims result in a number of investigative leads. Then subject matter experts and a health care investigative team are provided with the appropriate list of providers, and all their claims and procedures submitted during the time period evaluated. They are also given the reasons for selection for review, such as algorithm output and logic with

Model Prioritization and Development → Fraud Prevention System → Lead → Investigation → Action → Medicare Savings

FIGURE 4.1 Overview of the Fraud Prevention System. (From CMS. 2014. Report to Congress: fraud prevention system second implementation year. The Centers for Medicare & Medicaid Services. https://www. stopmedicarefraud.gov/fraud-rtc06 242014.pdf.)

data analysis summaries and key model attributes. Then, they decide which one of these leads warrant further investigation and potential action for recoupment. Further investigation is either done on the whole set of population, or on a statistically valid sample. For the selected set of claims, government contractors investigate the leads through boots-on-the-ground activities such as site visits, beneficiary interviews, and medical chart review.

These investigations can result in administrative actions including revocation of billing privileges, implementation of pre-payment review edits, referrals to law enforcement, and suspension of payments, as well as requests for further training and financial settlements. In particular, we are interested in the generation of leads using statistical methods as done by the CMS via the FPS.

Using the FPS, CMS has identified more than $1.5 billion in health care fraud, waste, and abuse within the Medicare Fee-for-Service program between 2011 and 2015. It has protected taxpayer dollars while reducing the burden on legitimate providers and suppliers and ensuring beneficiary access to necessary health care services. For instance, an ambulance company in Texas was found to be submitting claims for non-covered and not rendered services. A medical clinic in Arizona was billing excessive units of services per beneficiary per visit, which was found to include repeated and unnecessary treatments upon review. Both of these providers had their enrollment revoked.

Overall, the national return on investment was reported as $11.60 for each dollar spent on the FPS. The FPS is simply based on running many models concurrently. The CMS lists the main types of algorithms as predictive models, rule-based, outlier-based models, and network-based models. In this chapter, we present predictive methods in detail while the next chapter discusses algorithms such as outlier-based and network methods.

PREDICTIVE METHODS

Data analytics methods are classified with respect to the availability of labeled data. Labeled data refer to known values of the output variable of interest, such as the outcome of a claim, whether it is fraudulent or not. Using these labels, predictive methods are used to come up with predictions for new claims. Think of a black box of claims, all with different characteristics, and you know if they are fraudulent or not. When a new claim is submitted, you simply compare it with this black box of claims and predict how likely it is to be fraudulent. This setup is also called supervised learning, and these methods are called supervised methods. Supervised methods differ from each other by the way they summarize what is going on in the black box, whereas so-called unsupervised methods do not require labeled data and do not have particular output values. They aim to find hidden structure in data. In case of a limited amount of labeled data and/or expert opinion, both approaches can be combined within a hybrid approach. We will present unsupervised methods in detail in the next chapter, with a focus on the discovery of new fraud patterns.

Fraud detection has been a well-studied topic in many different domains such as credit card, telecommunications, auto claims, and computer security. The general objective is to reveal patterns and make predictions for the target variable. With respect to that, credit card fraud detection systems have been very successful. Analytics can be used in real time to stop potentially fraudulent transactions and request further verification before the payment confirmation. I am sure most of you have gotten a text or an email from your credit card company for verification. Sometimes, it may be a one-time large purchase you rarely make, but sometimes it is your credit card being used 1,500 kilometers (approximately 932 miles) away from you.

Health insurance claims can be argued to be similar to credit card transactions. Both systems record the information of the purchaser and vendor, and details of a particular transaction. Having said that, health care claims data bring unique challenges such as:

- Large number of parties involved in health systems and lack of communication across these organizations

- Complex nature of claims data

- Potentially poor data quality and incomplete data because of challenges with incorporating third-party data

- Constantly changing legal environments and medical policies

- Constantly changing tactics by fraudsters

Predictive models are frequently used to deal with such complex health care claims data. The way it works is similar to the computation of your credit score. The financial system uses your payment history, income, debts and types of credit to come up with a value that summarizes your creditworthiness. In health care fraud assessment, health care providers are similarly evaluated based on their claims and billing data, and a fraud risk score is computed. Usually, higher scores correspond to greater risk levels.

Fraud risk scores, or so-called fraud propensity scores, are computed for claims based on their likelihood of being fraudulent. The models use a variety of data sources including but not limited to the provider's past history, association with other providers and patient data. So, initially, there is no need to enter any rules, or decide on thresholds. You only need to provide data of audited claims, so-called labels. The more numerous and variable these labels are, the better the algorithm learns and the predictive accuracy increases.

Such predictive approaches provide information about the likelihood of claims to have payment issues based upon past data. This works fairly well in domains with a number of known fraud patterns. One can also use thresholds. Beyond a certain score, a provider can be flagged for further investigation.

Next, we describe more frequently used classes of predictive methods such as regression and classification for health care fraud detection.

PREDICTION OF OVERPAYMENT AMOUNT AND FRAUD PROBABILITY

Regression is among the most widely used statistical models. It can be used to explain the influence of a set of input variables on an output variable, and to predict the output variable. Most popular regression models are based on using a linear function, which corresponds to summarizing the relationships with a line. Such regression models have been applied extensively in many fields including finance, medicine, economics, production, retail, and social media.

A number of applications of regression models can be listed as:

- Predicting house prices using size, neighborhood, number of bedrooms, and market condition

- Forecasting demand as a function of weather, day of the week, season, item characteristics

- Modeling market share as a function of advertising expenditures, product quality, customer brand awareness, location, and competition

- Modeling stock prices as a function of company, industry, and market conditions

- Predicting number of likes or clicks on a particular post in social media using the recency, relevancy, and time of the post as well as the characteristics of the user

- Analyzing the impact of a treatment as a function of duration, frequency, and patient attributes

How can regression be used in health care fraud assessment? Regression can help auditors to predict the overpayment amount. Although ensuring the system is free of fraud is important, auditors tend to follow the big money transactions. Their time is limited and expensive, and they would rather reveal a $1 million scheme than a $10 one.

Let's assume that whistleblowers provided you with evidence that two providers may be involved with fraudulent schemes. However, you have resources to audit only one provider. You may predict the overpayment amounts for both, and decide to audit the one with the higher predicted overpayment amount. This can provide the auditor with an educated guess for an investigation lead. We should emphasize that the final decision of the legitimacy of the claim can only be made after the actual audit.

An option for regression modeling is to consider all the related variables in the data set to predict overpayment amount. You do not necessarily need to identify all of the variables to include in the model. Using a so-called forward or backward stepwise regression method, you can let the method choose the variables that provide the best fit and prediction.

Regression can also be used to predict the probability of fraud, also referred to as the fraud propensity score. Using a number of variables such as provider, patient, or claims characteristics, the fraud probability can be predicted to construct a fraud score of a given claim. Then the auditor can use the ranking of claims that presents how likely a given claim is to be fraudulent.

When the dependent variable is a categorical variable, like 0 or 1, the regression model is called logistic regression. For instance, instead of predicting the numerical value of an impact of a treatment, you may just want to predict if the patient survives or not. Variables with categorical outcomes are known as categorical variables. Such applications range from modeling the probability of default or fraud to predicting the churn rate of a customer or modeling the failure rate of a production system.

It should be emphasized that pre-processing, such as handling outliers as well as missing data, is paramount. In a logistic regression setup, the target (dependent) variable is whether or not the claim is fraudulent: 1 if yes and 0 otherwise. How can we predict if a new claim is fraudulent or not?

We simply look at all observations with fraudulent claims and aim to summarize those characteristics. In the data set, we would have a set of claims that are already audited. Simply, we know if they are fraudulent or not. For instance, suppose claims submitted for patients who are older than 75-years-old who get service more than 48 kilometers (approximately 30 miles) away from their residence have been fraudulent. Then, it is more likely that the predictive algorithm would flag a claim submitted for a 77-year-old who got service 80 kilometers (approximately 50 miles) from her residence. The probability of that claim being fraudulent would be high and it would be a candidate for further investigation.

However, that new claim may be totally legitimate. It may be the case that she got service while visiting her daughter.

Keep in mind that predictive algorithms are based on using the labels and the information in the black box of variables. So, they are based on the assumption that past activity is an indicator of future. Therefore, the labels and the information need to be frequently updated. For instance, in our example, this may warrant adding patients' children's residences into consideration to improve the model.

One issue with these claims data sets is that fraud is a relatively rare event. That being a problem sounds strange, does it not? Although it is good for the system, for modeling purposes having not too many fraudulent outcomes will make it harder to understand the characteristics that relate to fraud. To overcome such issues, relatively advanced pre-processing techniques such as oversampling can be utilized. We will not get into these details of these.

Logistic regression can also be considered as a classification algorithm for given threshold levels, which is to be discussed next.

CLASSIFICATION OF HEALTH CARE CLAIMS

Classification is the task of classifying a bunch of objects into separate groups using labeled data points. First, you identify "signatures" for each of these buckets. Then, you compare the new data and learned patterns, and assign the new observation to one of the groups. Examples of applications of classification algorithms include

- Finance: approval or denial of a loan application
- Marketing/sales: purchase of an item or not, response rates
- Text mining: a given speech supporting a given idea or not
- Computer information systems: an email being spam or not
- Medicine: diagnosis of a disease
- Video games: actions of a character given your commands and moves

In health care fraud assessment, classification type algorithms can identify the category of a new claim on the basis of labeled data with known category memberships. These algorithms mostly assume that the classes are not overlapping, and the attributes are independent given the class. Logistic regression is one of the classification algorithms that can predict the class of a categorical variable given a number of variables. The fraud probability of each claim can be compared with respect to thresholds, and auditors can investigate the claims that are higher than a user-defined threshold.

Decision trees are another widely used and easy to communicate classification algorithm. Decision trees are based on making a decision on the class of the claim using the values of a target variable. The list and order of variables are generally decided with respect to their information content. Decision trees have generic rules which are relatively easy to interpret especially with a small number of categories. One can choose a sequence of variables and their cut-off points to construct branches and classify new claims. The interpretability of decision trees may decrease with increasing size of data.

Let's discuss how decision tree output can be used for health care fraud assessment with an example output. Figure 4.2 presents the output of a decision tree for 2000 claims. This is retrieved using the statistical software tool R. Out of 2000 claims, 1789 of them are found not to be fraudulent. That is how the decision tree starts as can be seen at the top of the tree (no:1789/2000). Then, we use the variable "FraudConnections" for further classification. This variable measures whether that provider has any ties to providers and patients who were found to be fraudulent in previous audits. The outcomes listed as

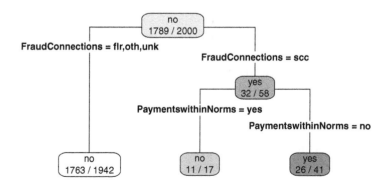

FIGURE 4.2   A decision tree output for classification of health care claims.

failure (flr), other (oth), and unknown (unk) correspond to lack of any fraudulent connection. As expected, most of such claims are classified to be not fraudulent (see the left branch of the decision tree, no:1763/1942).

On the right side of the decision tree, fraud connections are shown to be success (scc) for 58 claims. Out of these 58 claims, 32 of them are labeled as fraudulent. This provides us with the best indicator so far, with a 32/58 (55%) fraudulent claim proportion. Then, "PaymentswithinNorms" variable is used. This indicates if that provider is found to be involved with payments within norms with respect to his/her peers or not. When the outcome of this variable is "yes," as shown on the left branch, 11 out of 17 claims are found not to be fraudulent (no:11/17). Whereas for 41 cases, the provider was involved with out of norm payments, and 26 of them correspond to fraudulent transactions. So, this is another clue that can be used to determine which claims to audit.

Analyzing this decision tree, we can argue that providers with fraudulent connections and payments out of norms are more likely to be classified as fraudulent. Using this output, they are more likely to be audited.

Other classification algorithms used for health care fraud detection are relatively complex for the scope of this book. For instance, neural networks are complex pattern recognition and prediction algorithms that are inspired by the information webs among neurons within the biological nervous system. By construction, neural networks learn by using labeled data, and can handle complex, large data sets and non-linear variable relationships. Many layers can be constructed in neural networks, which are also referred to as deep learning methods. Other examples include Bayesian networks, k-nearest neighbor algorithm and support vector machines. Combinations of several supervised algorithms as an ensemble are widely used in industry to improve the classification performance. The application of more complex methods requires statistical expertise.

ACCURACY AND VALIDATION

Are the results of these well-established predictive methods generalizable? How do we know if they are accurate with a new data set? These remarks about accuracy and validation are very important.

Back to the example introduced at the beginning of this chapter: let's assume this auditor in Texas chose and used a predictive method with success. Can she keep using it for the foreseeable future? The success of predictive methods lies in the accuracy of the labels in representing fraud patterns. While validating predictive methods, the data set is generally separated into two: so-called training and test data sets. The training data are used to form the model in the first phase, say with 80% of the data. The remaining 20% of the data are kept separately from the training process and are treated as if they were future data. So, we can test the success of the predictive method. This process is repeated a number of times with different training/test data combinations. So-called cross-validation balances the impact of randomness while assessing the prediction accuracy.

The reproducibility of the success of the algorithms is a major concern, especially if the pattern of the fraud changes. More labeled data are generally needed to update the algorithm. However, every labeled data point means that we need to have already audited enough respective claims. Since fraud is a rare event, it can be expensive and time-consuming to get a big enough sample size of fraudulent cases. Even if one has access to the resources, it may be challenging to discover new fraud patterns.

Sometimes, the tool may work well for the training data but results in larger errors for test data. This is especially prominent with skewed data such as health care claims, which have many more legitimate cases than fraudulent. There are ways to overcome so-called overfitting, but it is too advanced to discuss in this book. Let's continue with an example of how to assess accuracy.

Let's say an audit contractor investigates the effectiveness of the tool for 300 randomly picked claims. The predictive method is found to predict 287 of the claims correctly including 35 overpaid claims. Whereas it incorrectly predicts three actually legitimate claims as

TABLE 4.1   Confusion Matrix for a Predictive Method

	Actual Legitimate	Actual Overpayment
Predicted Legitimate	252	10
Predicted Overpayment	3	35

overpaid, so-called false positives. And ten of the actual overpayments are missed by the tool and predicted as legitimate. See Table 4.1 that summarizes these outcomes, also known as a confusion matrix.

What does this table really correspond to? Three providers had to go through an investigation although they had done nothing wrong. These investigations cost money and time, for both auditors and also these providers. Eventually, these mistakes decrease the trust of providers in the system.

Whereas ten claims were fraudulent, and could not be detected by this algorithm. That is one of the Achilles' heels of the predictive methods that work with labeled data. They would not be aware of a fraud pattern unless labeled data about it become available, which in turn will not be available, since the system thinks those claims are legitimate and not worth investigation. There are methods to break such vicious cycles and reveal emerging novel fraud patterns; these will be discussed in the next chapter.

DISCUSSION

Now we have discussed a number of predictive methods, the question becomes how to choose the proper fraud analytics tool. Well, it depends. There is not a clear-cut answer, but some aspects are more important than others when making a decision.

Investigative resources are almost always limited to examine all suspicious activities. Prioritization of leads with respect to financial recovery potential and routing potentially fraudulent cases to appropriate team members with relevant expertise are crucial. Use of predictive algorithms to prioritize highly probable fraud cases results in potentially more and easier recovery instead of manual prioritization. For instance, using a predictive algorithm output, one can decide to investigate the claims with high-risk probability scores, high payment values, and highly defendable characteristics such as fraud score and related reason descriptions.

For predictive methods, continuous testing and validation of the model are necessary. The assumptions behind these approaches are important and need to be communicated with the auditor who makes the final decision based on the outputs. The ease of use and visualization via dashboards, graphics, and maps that assist the user are also crucial.

At times, so-called ensemble approaches give the chance to use these models for different portions of the data and combine the output. This may improve the unbiasedness of the algorithms and decrease false positive ratios.

Actually, having too many false leads (positives) is an often-seen problem. Incorporation of relationships retrieved via complementary data sources to reveal previously unknown relationships can help with that. Frequent updates of rules and training labeled data can be conducted by scheduling more frequent software and data checks. In addition, labels should be recollected whenever there are changes in legislation and prescription patterns.

Data coming from many different sources emphasize the importance of data warehousing and consolidation of data well as data pre-processing and quality checks. Repeat offenders can be identified and insider or collusive fraud can be uncovered by incorporating all available data such as public records and social media. Successful use of accurate predictive methods is and will be paramount to protect taxpayers' money. However, it may be expensive to maintain the accuracy of these methods; therefore, they may not be very suitable for detecting emerging patterns. The next chapter presents methods to handle novel patterns of fraud.

KEY TAKEAWAYS

1. Data analytics methods can help health care auditors to detect potentially fraudulent claims pre-payment and post-payment.

2. CMS reported more than $1.5 billion in savings due to analytics initiatives with the Fraud Prevention System.

3. When there are labeled data, supervised methods such as regression and classification can be used to predict the overpayment amount and fraud likelihood of a particular claim.

4. Classification methods can be used to predict classes of health care claims.

5. Accuracy of the predictive algorithms can be assessed by using validation methods.

6. Labels need to be updated frequently to enable predictive algorithms to adapt to changes in fraud patterns.

ADDITIONAL RESOURCES

1. Data mining textbooks:

   a. James, G., Witten, D., Hastie,T, Tibshirani, R. (2013) *An Introduction to Statistical Learning.* Springer. Available at www-bcf.usc.edu/~gareth/ISL/.

   b. Hastie,T, Tibshirani, R., Friedman, J. (2009) *The Elements of Statistical Learning: Data Mining, Inference, and Prediction. Second Edition.* Springer. https://web.stanford.edu/~hastie/Papers/ESLII.pdf.

   c. Shmueli, G., Bruce, P. C., Yahav, I., Patel, N. R., Lichtendahl Jr, K. C. (2017). Data Mining for Business Analytics: Concepts, Techniques, and Applications in R. John Wiley & Sons, New York, NY.

2. Relevant news article:

   Agrawal, S., W. R. and Bowman, K. (2016). Medicare big data tools fight and prevent fraud to yield over 1.5 billion in savings. MMAP. https://bit.ly/2LkIVNe/

3. Relevant governmental report:

CMS (2014). Report to Congress fraud prevention system second implementation year. The Centers for Medicare & Medicaid Services. https://www.stopmedicarefraud.gov/f raud-rtc06242014.pdf.

4. Scholarly articles about use of predictive data mining algorithms for health care fraud detection

a. Bauder, R., Khoshgoftaar, T. M., and Seliya, N. (2017). A survey on the state of healthcare upcoding fraud analysis and detection. *Health Services and Outcomes Research Methodology*, 17(1):31–55.

b. Dash, M. and Liu, H. (1997). Feature selection for classification. *Intelligent data analysis*, 1(1–4):131–56.

c. Dionne, G., Giuliano, F., and Picard, P. (2009). Optimal auditing with scoring: Theory and application to insurance fraud. *Management Science*, 55(1):58–70.

d. Ekin, T., Ieva, F., Ruggeri, F., Soyer, R. (2018) statistical medical fraud assessment: Exposition to an emerging field. *International Statistical Review*. 86(3), 379–402.

e. He, H., Wang, J., Graco, W., and Hawkins, S. (1997). Application of neural networks to detection of medical fraud. *Expert Systems with Applications*, 13(4):329–36.

f. Li, J., Huang, K.-Y., Jin, J., and Shi, J. (2008). A survey on statistical methods for healthcare fraud detection. *Health Care Management Science*, 11:275–87.

g. Shin, H., Park, H., Lee, J., and Jhee, W. (2012). A scoring model to detect abusive billing patterns in health insurance claims. *Expert Systems with Applications*, 39(8):7441–50.

CHAPTER 5

Discovery of New Fraud Patterns

OVERVIEW

Rockwall is a small city in Texas, near Dallas. It was an ordinary small town, until the national news broke about one of its respected physicians. Dr. Jacques Roy was noticed to submit by far the most Medicare claims in the nation for home health services. Further investigation showed that Roy's office handled more home health care visits from January 2006 to November 2011 than any other physician's office in the country. As it was put by an Assistant U.S. Attorney during the trial, "A doctor cannot care for 11,000 patients at once."

The more his activities were investigated, the more dirt came out. Auditors found out falsified visits for un-rendered services that were not necessary to start with. People from homeless shelters were taken to lunch in exchange for allowing their Medicare information being used. Dr. Roy was eventually sentenced to 35 years in federal prison and $373,331 in restitution for health care fraud and false statements. His lawyers continued to suggest he was just a hardworking doctor who did what was necessary for his patients.

In reality, when agents searched his home in 2011, they found books about vanishing without a trace and hiding money in off-shore bank accounts. Agents also found two different fake identities with Roy's photo along with passport documents from Canada, where he was born. According to federal prosecutors, if he had pulled off his escape plan, Dr. Roy might have been living in Canada or France now under the alias of Michel Poulin. The investigators got to Dr. Roy early enough, so he could not manage to run away from prosecution. However, most fraudsters manage to pull out an escape, while all taxpayer dollars vanish and become almost impossible to recover.

Dr. Roy's extraordinary activities were revealed with the help of an outlier detection tool. How do these methods work? Why did Jacques Roy go unnoticed for at least five years? How can we reveal new fraud patterns and recover the overpayments before it is too late?

Predictive methods rank claims with respect to their likelihood of fraud, using known fraud patterns. But what about new patterns of fraud? If you were a fraudster, would you continue committing fraud the same way or mix and match your ways a little bit every time?

Fraudsters adapt to investigation outcomes as well as to changes in health care policies, and, therefore, the nature of fraud patterns change. Predictive methods are not able to capture changing fraud patterns unless they are frequently updated with labeled data. Such frequent updating can be expensive since retrieving labeled data in the context of health care fraud detection corresponds to the need for an actual audit, which can also be time-consuming. This lack of adaptability and the need for constant tuning have increased the attention on outlier detection, clustering, and association methods. These are also referred to as unsupervised methods—since they do not require labeled data.

These methods generally serve as pre-screen filters that list potentially fraudulent claims before the actual audit. This initial screening can decrease personnel costs as fewer transactions are reviewed. They are not dependent on a particular labeled data set. That is why they can be used to help detect changing fraud patterns.

Outlier detection methods are used to reveal activities which are abnormal compared to average or expected behavior. Clustering is the most common segmentation method and is based on grouping similar objects into clusters. These findings can be later used along with predictive methods including classification. Association methods, link analysis, and network-based methods study attributes that go together in order to discover hidden relationships in a large data set. This chapter introduces these methods which can enable auditors to detect dynamic fraud patterns.

The questions to be addressed include

- How can we detect excessive or very different claim submissions, using outlier detection methods?

- How can we group claims?

- How can association type algorithms be used to find links among claims?

- What are the aspects to consider while assessing the effectiveness of the analytical methods?

- What should auditors consider during the deployment of the statistical methods?

OUTLIER DETECTION: FINDING EXCESSIVE BILLINGS

Have you ever heard about the importance of dental braces in Texas? Between 2005 and 2010, Texas Medicaid has spent more money on braces than the other 49 states combined. Given that Texas state population is less than 10% of the population of the United States, it is surprising. Either patients take really good care of their teeth in Texas compared to the nation, or providers bill excessively.

Audits revealed that doctors have routinely billed Texas Medicaid for uncovered procedures. These include putting braces on youngsters for purely cosmetic reasons and performing unnecessary root canals on small children.

Some of these unnecessary procedures even resulted in the deaths of pediatric patients. Extra scrutiny is good to ensure patient health is protected and taxpayer money is not wasted. However, extra audits also prompted some dental offices to stop serving Medicaid patients. So, health care fraud eventually hurts legitimate providers and deserving patients, one way or another.

Durable medical equipment, such as wheelchairs, scooters, and walkers, increases quality of life for many senior citizens. However, this equipment is also abused a lot. For instance, there were instances of billing everyone in a particular nursing home, as if all residents needed and received a wheelchair.

Seniors are especially targeted in such schemes. It is illegal for a medical supplier to make unsolicited telephone calls to people with Medicare, other than consented to and follow-up calls. Some suppliers get around this by hiring independent marketing firms. These high utilization rates can be identified by outlier detection methods.

A common aspect of these examples is excessive activity. How do we define excessive? This is the tricky part because of the subjective nature of health care and treatments. One specialist may be legitimately billing for hundreds of thousands of dollars for a patient who is getting a specialized cancer treatment, whereas the activity of a hundred-dollar billing may be excessive if the diagnosis only requires a ten dollar procedure. So, it is important to compare apples to apples, although the complexity of health care makes this challenging.

There are a number of ways to define normal activities. For each set of providers or procedures, there may be benchmarks that define normal ranges and acceptable thresholds. Then, claims that exceed a certain dollar amount or providers that overcharge for a given service can be flagged.

Outliers are defined as out of the norm, abnormal, and, therefore, exceeding the thresholds. The biggest challenge is to determine what to measure and the respective threshold levels. Setting thresholds or deciding the extent of normal require subject domain expertise.

If you set the thresholds too high, then you may miss the fraudulent activities. If you set it too low, then there will be too many claims to investigate. Integrated adaptive outlier detection methods may help to detect excessive billings, even before the payment. The bottom line is that labeled data may not be needed to reveal new or previously unknown patterns of fraud.

Some approaches analyze the data of billings over time and aim to detect spikes and billings outside of the trend line. For instance, a primary care physician may see more patients during flu season, and claims will spike. That would be deemed as normal. Whereas spikes occurring at odd times for that provider can be flagged for further audits. Sudden increases in billings or number of patients also trigger investigations. Even a provider suddenly starting to bill for very old patients can be noteworthy and may warrant assessment.

In order to compare apples to apples, we need to work with homogenous groups, or so-called peer groups. These peer groups are set by the characteristics of the providers, claims, or patients. For instance, the billings of a particular provider and the billings of providers in the same specialty can be compared. Then it can be argued that the providers who bill for more services than 99% of similar providers are suspicious and should lead to further investigations.

An alternative is to examine the overall activity, and aim to come up with homogenous groups. We will present a more detailed discussion on those grouping activities later in this chapter.

Descriptive statistical and basic visualization methods, which are discussed in Chapter 2, can be used as simple outlier detection methods. For instance, providers' billings for a given procedure can be compared using a boxplot. These can result in audit lists such as:

- Doctors who treated more patients a day than their peers
- Providers administering far higher rates of tests than their peers
- Providers billing far more costly claims, on a per patient basis, than their peers
- Providers with a higher ratio of distance patients than their peers
- Providers that prescribe certain drugs at a higher rate than their peers

What if we want to compare the provider's billing behaviors for more than one procedure? One option is to use a composite ranking. You can score the difference from the expected value, or average, for each procedure, and come up with an aggregate measure.

An alternative approach could be to use concentration functions and Lorenz curves. This can help in investigating the billing differences of a provider among a variety of prescribed services. The main focus is to compare each provider's billing behavior to the average behavior in their peer group. Selecting that peer group becomes even more important.

The peer groups are selected using assumptions such as a group of providers having similar characteristics based on provider specialty, region, and number of years of experience. Providers within peer groups are assumed to be providing similar services to similar patient populations. The distributions of these billings can be compared using a similarity measure. We will use the Lorenz measure in the following example.

Let's assume we have cardiologists billing for a total of 100 procedure codes in a homogenous region. We are interested if any of them have very different billing patterns compared to the benchmark—assumed to be average in this case. We compute the cumulative billing percentages for all procedures. The plot of these cumulative billing percentages versus the cumulative percentage of the corresponding procedure is referred to as the Lorenz curve. If a provider exactly mimics the average billing behavior, then his/her line will be on top of the benchmark line. Here is an example that compares two providers (see Figure 5.1). The provider that is denoted by the dashed line is more similar to the benchmark (bold line) compared to the other provider (dotted line).

Once peer groups and benchmarks are determined, statistical measures and visualization tools can be used to display outliers. This is one of the many methods to detect outliers. Please see the end of this chapter for references of more applications.

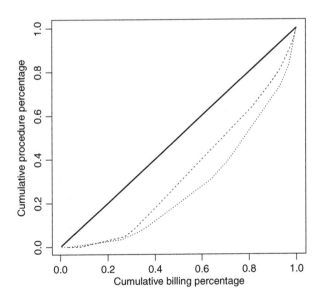

FIGURE 5.1　Lorenz curves for two providers (dashed and dotted lines) compared to the benchmark (bold line).

CLUSTERING: GROUPING HEALTH CARE CLAIMS

The Office of Inspector General (OIG) often receives calls from whistleblowers about health care fraud. Sometimes, it is a whistle-blower suggesting that one hospital has billed Medicare for millions of dollars' worth of up-coded (overcharged) services. The auditors aim to act on the provided information, but many hospitals bill more than a million claims every year. How do you decide if these claims are up-coded or not?

The audited item can be compared with identical claims or providers, the so-called peer groups. While presenting outlier detection methods, we discussed the importance of finding peer groups. At times, we know what the peer group is, such as the specialty of the provider. For example, Berenson–Eggers Type of Service (BETOS) categories can be used to decide on peer groups for providers. BETOS codes are assigned to every procedure and consist of readily understood clinical categories. They are stable over time.

As an alternative, further data such as patient profiles for each provider can be utilized to construct sub-peer groups. That is where clustering can be beneficial.

Clustering is simply used for grouping, or so-called segmentation. The objective is to find groups, aka clusters, that have similar observations within the group, and are as different as possible from other groups. Let's say you have 100,000 movie goers who may be fond of different genres of movies. You have access to their characteristics, such as age, time spent on social media, and favorite TV shows. But you do not know their favorite genre(s).

Clustering can help group them based on the type of movie they may like. Ideally, when you group these moviegoers, you would put all action genre lovers into the same group, while drama and comedy moviegoers would be in separate clusters. Keep in mind you do not really know their favorite genre, so you try to group them with respect to other characteristics. Given these attributes, the objective is to find clusters such that data points in one cluster are more similar to one another whereas data points in separate clusters are less similar to one another.

Once the clusters are identified, an option is to apply labels to each cluster to classify each group based on its characteristics. These findings can be later used to build predictive models including classification or decision trees.

How can we use such an approach for health care fraud detection? Let's say we have all the claims submitted by a hospital. We can aim to find out unusual or outlying clusters that cannot be explained by "normal" claim behavior.

Clusters can also be used as the first step of any statistical analysis to ensure the group of interest is homogenous. For instance, in U.S. Medicare, socio-demographic and location variables are used to cluster geographical regions before analyzing billing behavior. Another way is to form physician practice profiles with respect to number of visits, prescription percentages, and expenditures.

Most of these methods are so-called hard clustering approaches. A movie fan can be assigned to only one group, suggesting she can only like one type of movie. One provider can only correspond to one group at a time. However, the real world does not work that way. Therefore, so-called soft, aka probabilistic clustering methods are proposed. Each provider is assumed to have a certain probability of billing to any given cluster. This can be especially beneficial for providers who work with different sets of patients that require different levels of procedures. One provider may act like a cardiologist 50% of the time while billing like an internal medicine specialist 40% of the time. So-called Bayesian methods can deal with uncertainty by assigning probability distributions to all unknown events and variables.

Some of the relevant methods also help you do both grouping as well as outlier detection, which are referred to as Bayesian hierarchical methods, but we will not get into detail in this book. Hierarchical structure helps with dealing with complex data which can be modeled in layers. This can help you reveal the hidden patterns among providers and medical procedures which would otherwise be missed within a big data set.

Let's present how this can be applied to health care claims data. Figure 5.2 presents an overview of the data set of interest, which

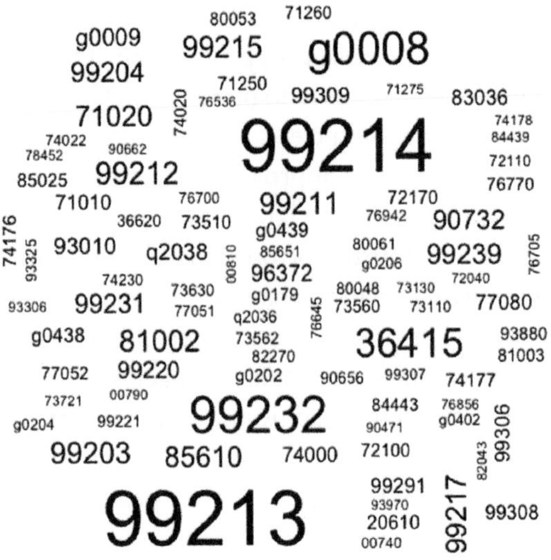

FIGURE 5.2   Health care procedure cloud.

is a collection of Medicare Part B claims. This is also referred to as a health care procedure cloud. This is very similar to so-called word clouds that are used in text mining. This procedure cloud lists the procedures on a map depending on their relationship. If two procedures are billed by similar doctors, they would be more closely located on the map. The size of each procedure represents the frequency of billings for that procedure. For instance, 99213 is a frequently billed procedure, therefore it is larger in the map. You can use this as a pre-screening mechanism within your pre-payment or post-payment analytics framework.

A Bayesian hierarchical method can be used to understand which procedures are frequently billed together. This is a soft-clustering method, so each procedure is a member of all groups (clusters) with a certain probability. In text mining, these groups are also called topics. Table 5.1 provides an output of this method. It shows the five most frequently billed procedures in a particular group with their descriptions and billing frequencies. What do you recognize when you examine this group?

TABLE 5.1   List of the Most Frequently Billed Procedures in a Descending Order

HCPCS Code	Description	Frequency
92012	Eye exam established patient	0.324
92014	Eye exam and treatment	0.156
92083	Visual field examination(s)	0.075
S0620	Routine ophthalmic exam including refraction—new patient	0.062
92134	Computerized ophthalmic imaging retina	0.051

Even for our untrained eyes, the answer can be relatively simple. You may notice that a collection of eye-related procedures is displayed in Table 5.1. But how can this be used for audits?

You would expect ophthalmologists and optometrists to bill for this group and these five procedures. First, we can check if that is the case. In addition, a follow-up investigation would include specialty providers other than eye doctors billing for this group the most. Or we can investigate the ophthalmologists and optometrists who do not bill for this topic the most.

As you can recognize, in health care, fraud assessment tools do not give you ultimate answers. Most often than not, they result in more questions. It is all about funneling your energy and resources to potential problematic areas. Eventually, an audit would tell you whether the investigated activities are fraudulent or not.

Once you identify a suspicious provider, more statistics of that provider, such as its proportion of billing to each group, can be used. That helps to understand if that physician is an outlier or bills similar to his/her peers. A variety of descriptive statistical methods are already presented in Chapter 2. Such visual evidence helps to see how different that unusual provider is. In Figure 5.3, a boxplot displays the proportions of billing of each doctor in a peer group (referred to as theta in the figure) to four groups (referred to as topics or clusters in the figure). The unusual provider is denoted with a triangle. You can see the differences in the billing pattern of that outlier doctor. For instance, the outlier doctor bills most

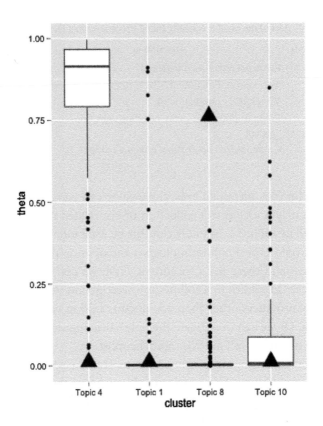

FIGURE 5.3   Select group proportions for the peer group.

frequently for the procedures in Group (Topic) 8. Whereas all his/ her peers most often bill for Group (Topic) 4.

Similarity computation tools used with clustering output can help identify potential teamwork as well as outliers. For instance, you expect a certain kind of cardiologist to have similar billing behavior, and therefore to have similar scores. A cardiologist with a high dissimilarity score compared to other cardiologists can be red-flagged. You may also have an anesthesiologist who bills similarly to these cardiologists. Such billing patterns can be a sign of collaboration or fraud. However, we cannot know that for sure before an actual investigation.

So far, we have discussed methods of grouping with respect to one aspect or one variable. Relatively new approaches are proposed to deal with dyadic data using so-called co-clusters. These aim to link two group types. For instance, the frequency of visits of groups of beneficiaries to groups of providers can be modeled with a co-clustering algorithm. Cardiologists may be found to serve older patients, forming a physician-patient co-cluster. If a particular cardiologist has been serving patients outside of his/her co-cluster, that may signal unusual activity. These methods can potentially reveal an emerging type of fraud called "conspiracy fraud" that involves attributes of more than one party of the health care system.

ASSOCIATION: FINDING LINKS AMONG CLAIMS

Which procedures are billed in a similar fashion? Which prescriptions are written by what kind of doctors? Can we find any potential associations and links among them? Association type algorithms are designed to help with these questions. This class of descriptive methods studies attributes that go together to discover relationships hidden in large data sets.

One of the early and popular applications has been to investigate customer behavior in retail stores. Therefore, this is also referred to as market basket analysis.

Social media companies use such algorithms extensively to understand user behaviors in their networks. In the auto insurance industry, so-called crash-for-cash schemes are uncovered using network algorithms.

The number and quality of relationships within a provider network can be analyzed in a similar setup. These algorithms may help reveal relationships, links and hidden patterns of information sharing and interactions within potentially fraudulent groups of providers and patients. This may be through links between businesses such as contact information, locations, service providers, assets, associates, or the relationships with a group of beneficiaries who are known to be within the fraudulent circle. Financial statements, private commercial records, property tax records, banking records, and voter records can all help reveal further connections. You can even consider text data and social media data to reveal relationships.

Association algorithms can reveal the connection between pharmacies that were owned by the same people and providers who share a large percentage of the same patients. These algorithms can also compute the similarity between providers based on activity and the likelihood of any given patient to get services from a provider. For example, you can expect any two providers with a shared specialty to bill for relatively similar procedures. If their billing records are too different, that may signal suspicious activity.

In July 2017, the Justice Department announced charges for 412 people, including many doctors, in a $1.3 billion health care fraud. Some of the fraud schemes were committed within a network and were very sophisticated. For instance, one of those consisted of a guy who brought in patients through staged accidents, a lawyer that represented the patients against insurance companies, a few doctors who billed for the services, and the leader who ran the clinic and took care of the financials. It takes a lot of time and effort to get to the leader of such a network scam.

Let's present a simple demonstration of suspicious provider networks based on their information. Drs. Lecter, Decimus, and Lebowski may have totally different billings and may be serving different sets of beneficiaries. But Dr. Decimus and Dr. Lebowski work for Roman Hospital, while Dr. Lecter works for Lecter Associates. On their websites, Lecter Associates and Roman Hospital seem to use the same toll-free number. In addition, both these providers excessively prescribe home health devices that are filled by Durden Durable Equipment Corporation. Paying attention to details can reveal these connections, and coming up with a simple network may reveal sham companies and potential fraudulent links.

In one case, the owner of a drug-treatment center in Delray Beach, Florida, recruited addicts to aid him in his schemes. His team attended Alcoholics Anonymous meetings and visited "crack motels" to persuade people to be "treated" by him. Offered kickbacks include gift cards, plane tickets, trips to casinos and strip clubs as well as drugs. This drug-treatment center was eventually charged with fraudulently billing insurance companies for more than $50 million for false treatment and urine tests over nearly five years.

The output of these link analysis tools can be displayed to show connections that otherwise would not be detected. Such tools can also be used to understand which drugs are prescribed together. Auditors can explore the existence and strength of links of drugs as well as their frequency of being prescribed by the same or similar providers.

These methods are valuable since a significant amount of fraud, waste, and abuse are found to be committed by organized, sophisticated, and collusive networks of providers and patients. It is more likely that fraudulent people are connected to other fraudulent people. A large number of overlapping patients would raise a red flag for further investigation. These may be fraud, but may also signal complementary services. This cannot be known without an actual audit.

Thresholds can be used to decide on the exceeded frequency level required to start investigation. The frequency of connections between certain types of entities, if they are much greater than normally expected, can be red-flagged. This also can help catch ghost patient billing. There are many unsuspecting patients out there whose identities are used to bill for procedures that are never provided.

Fraud patterns change. The networks should be continuously updated with new data to reflect these changes. They should keep evolving. One should be aware that they are very informative, but these tools take time and infrastructure to build in the first place.

EFFECTIVENESS OF THE ANALYTICAL METHODS

So far, we have presented a variety of analytical methods. But how effective are they? Let's say you have a pulmonologist who specializes in the surgical treatment of lung issues, and he is the only expert who can conduct that expert level procedure in a geographical region of ten million people. All such patients would come to him, and his billings would be very different to the other pulmonologists. Most outlier detection methods may flag him as an outlier and may recommend an audit. After the audit, this may be seen as a case of method going wrong. However, keep in mind this billing could also have been a result of someone overcharging and billing for unnecessary expensive service codes.

That is why integrated approaches that also use other information, such as clinical data or demographics, are very valuable. Integration of field intelligence, policy knowledge, and clinical expertise into the development of algorithms and evaluation of outputs can help retrieve actionable output. This can reduce false positives and boost efficiency by preventing big losses early and potentially in the pre-payment phase.

Unsupervised methods give us the unique chance of understanding changes in billing behaviors, health conditions, or relationships within the provider network. For instance, network analysis can capture the effect of a new hospital to the other providers in the region. However, for the maximum benefit, we need to consider clinical knowledge for the evaluation of the output. The context of the patient, provider, and their joint history are all important. Evaluation of risk profiles and fraud scores of individuals based on an ensemble of models is crucial.

Measuring the success of statistical methods has been a well-documented challenge. There are not any industry-accepted standard methodologies to calculate savings of predictive analytics technologies for health care program integrity. In order to assess the overall impact, one needs to consider:

- Direct fiscal impact
  - Potential savings from denial of claims based on pre-payment reviews and auto-denial edits
  - Avoiding costs by more thorough screening of providers and revocation of billing privileges and payment suspensions
  - Overpayment recoveries by audits and fraud investigations
  - Contractor costs
- Improvements in fraud, waste, and abuse prevention, detection, and deterrence
  - Removal of systemic vulnerabilities
  - Sentinel effects such as providers self-modifying their behavior so as to not commit fraud, waste, and abuse
  - Greater ability to validate and compare claims across different provider types
  - Optimization of existing staff resources through better and more focused data retrieval
  - Communication of the policy changes to close vulnerability gaps and prevent future risks

In their annual report, the CMS came up with a novel measure to assess the performance of health integrity efforts. The CMS measure of adjusted savings only considers the direct savings, but not the avoidance of ineligible payments. It should be noted that the potential savings due to changes in provider billing behavior are difficult to measure.

Measuring the effectiveness of models can also help us to understand when a model becomes outdated and needs updating or replacing. In addition, it helps to assess the return on investments. The methods may be working fine and may be generating many accurate leads. However, if they do not translate to action such as overpayment recovery that results in savings, their use may not be justified.

This brings us to the next problem. After having justified the use of an algorithm, how do the auditors deploy it?

DEPLOYMENT VIA RULES

How do we deploy these methods within an audit framework? Most often, they are embedded within a rule-based setup to filter fraudulent claims and behaviors and to discover emerging new fraud patterns.

Rule-based frameworks have many sources of information such as whistleblower tips. The rules can reveal providers who are not billing for legitimate beneficiaries or can detect geographically impossible billings. For instance, if a beneficiary is billed in Texas earlier that day, another claim that is submitted for him in Florida one hour later has a high potential of being fraudulent.

You can also flag providers who bill for stolen Medicare identification numbers with a simple rule. Such applications are also referred to as identity analytics that address eligibility fraud. These can be performed by standard enrollment qualification processes that include verification and authentication of all provider and beneficiary identities. Identity analytics can help eliminate unqualified and risky providers from the network. It should be noted that such rules can also be used to reveal the existence of known schemes and obvious patterns that are retrieved based on retrospective reviews and descriptive statistical analysis.

Although they can be simple and effective for fraud detection, rule-based methods have a few disadvantages. First, it is time-consuming to construct rules at first. Then, subject matter experts need to update the existing rules with respect to changing fraud patterns. Otherwise, poor maintenance would result in false positives and a diminishing number of recoveries. In addition, new fraudulent schemes would remain uncovered, so-called false negatives. But once you create the rules, they easily generate leads. Sophisticated analytical skills are not required; anyone can use these after an initial short training.

On the other hand, fraudsters can game these algorithms with information about the rules and their thresholds. For instance, let's suppose, claims worth under $100 are only investigated if the overall billing activity of that provider is excessive compared to his/her peer doctors. If the fraudster has this information, he/she can get away with billing for fabricated claims each for $99.90 with a total number less than the threshold; so that will not be deemed as excessive.

That is why it is important to use outlier detection in conjunction with other methods. That may result in decreased false leads. In order to customize methods, it is a good idea to involve the investigators in the process. A close cooperation between physicians, statisticians and policymakers would be very beneficial during the stages of defining and tuning the model as well as analyzing and interpreting the results. Use of medical knowledge is expected to improve performance, despite adding additional cost and complexity.

Hybrid approaches can improve the deployment performance of health care fraud detection methods. For instance, the CMS has been continuously refining their existing models based on the feedback received through the FPS and insights from field investigators, policy experts, clinicians, and data analysts. Ensemble algorithms combine the strengths of the supervised models for certain aspects of the problem and help find the outliers to improve the predictions. Unsupervised methods can improve the performance of supervised methods. For example, a clustering algorithm can be applied to divide all insurance subscriber profiles into groups. Then a decision tree can be built for each group and converted into a set of rules. Clustering can overcome the deficiencies of decision trees with larger data sets and many categories. When an unsupervised method is followed by a supervised method, the objective is usually to discover knowledge in a hierarchical way.

CURRENT EFFORTS

Implementation of effective data analytics programs offers several advantages, including a positive return on investment that can exceed that of traditional methods and strengthen program integrity safeguards. Pre-payment analytics help with early detection of improper payments. Recognized fraud patterns can be used to improve the initial review of claims. The potential areas of further losses can be identified, and the necessary actions such as closing policy loopholes can be taken.

Identification of predictors of improper billing can also lead to the development of new and more effective models for post-payment audits and recoveries. Predictive analytics coupled with unsupervised methods can help with comparing providers of the same type and identifying long-term trends. Then these trends can be used to build novel pre-payment rules to check for future claim submissions.

The CMS reported total savings of more than $39 billion of improper payments in Medicare during 2013 and 2014 (CMS, 2016b). The related efforts include training programs, avoidance of incorrect submissions via improved billing systems and better pre-payment, identity controls and actual overpayment recovery.

A recent news release (CMS, 2016b) lists recovery tools as:

- Enforcement of the False Claims Act by the Department of Justice (DOJ)

- Efforts of the Medicare Fraud Strike Force, which consists of OIG and DOJ members, against organized crime

- Use of advanced fraud detection technologies by OIG and CMS

- Enhanced provider screening and enrollment requirements set by the CMS

- Increased collaboration among the DOJ, OIG, FBI, and CMS via the Health Care Fraud Prevention Partnership

- Senior Medicare patrols by groups of volunteers who educate and empower their peers to identify, prevent, and report health care fraud

- Training and education programs for users such as the "Help Prevent Fraud" campaign

When it comes to application of statistical methods, each method has its own particular advantages and limitations. Relatively more complex methods such as network analysis may be more time-consuming and are complicated to construct at first. However, they generally provide a comprehensive understanding of health care billing patterns, which may result in more accurate leads. Table 5.2 summarizes the methods that are discussed in Chapters 4 and 5 with major application areas as well as their main drawbacks.

Fraud detection schemes will keep evolving. Technological developments also help fraudsters to continually become more inventive and resourceful. Fraudsters are flexible enough to change their tactics based on their knowledge of the government's focus. Therefore, there will not be a bulletproof fraud or abuse detection technique.

Simultaneous and concurrent use of multiple methods offers the best chance for detecting both opportunistic and organized fraud. This can help in adapting to ever-changing fraud and abuse schemes, successful pre-payment detection, and the ability to deal with billings that are in large amounts and variety. Having

TABLE 5.2   Summary of Main Fraud Detection Methods and Applications

Method	Application	Drawback
Outlier Detection	Detection of different behavior	Need for well-defined peer groups
Regression	Overpayment prediction	Need for labeled data
Classification	Grouping with respect to fraud type	Need for labeled data
Network	Revealing organized crime rings	High initial investment
Clustering	Grouping claims	Need for clinical insights

a data management and analytics plan in advance can help with the objective of revealing high-risk relationships using all patient, provider, and claims data.

The utilization of better identity checks and pre-payment analytics are expected to help identify claims that are not eligible. Quick processing of post-payment analytics tools can identify leads for early intervention by Medicare administrative contractors. Ongoing projects of the CMS include the integration of the FPS with the claims-processing system to enable claims denials or rejections directly through the FPS. This was successfully piloted for the rejection of certain physician claims. The CMS also launched a pilot project to explore opportunities to leverage other interventions for resolving leads in the FPS, such as provider education or medical review.

Statistical methods have enhanced the data analysis capabilities of health care fraud investigation teams. The interfaces and visual aids facilitate data summary and analysis, especially in the cases of comparisons. Such integrated capabilities allow investigators to capture all findings that are relevant to an investigation, including claims data, network diagrams, case notes, surveillance video, and any other external structured or unstructured data. Combining the auditor expertise, data analysts, as well as the traditional investigation skills, is crucial for the battle against fraud.

For the ideal implementation of health care fraud analytics, an integrated system which can combine a variety of data sources and expert knowledge such as medical and clinical insights, financial analysis, death/birth, and prison records is necessary. Such data resources help algorithms learn and improve themselves. This is crucial especially for proactive detection against newly evolving fraud patterns. For instance, clinical feedback may be the main difference maker for the success of an anomaly detection algorithm. The algorithm may return unusual activities as statistical outliers and flag many specialists such as oncologists and cardiothoracic surgeons. However, medical experts can help prioritize leads, and argue that sicker patients may visit with that provider because of the expertise of the physicians and hospital. Clinical insights and medical logic can separate the medically needed and legitimate billings. In order to address this, there are ongoing efforts to use artificial intelligence-based tools within the health care domain.

In addition, there are a number of promising methodological advancements in recent years. More advanced time-trend analysis-based tools have the potential to become more widely used. These can show the billing development of networks or regions over a time horizon. They can also help tune other models and update thresholds in rule-based methods. Analysts may also forecast the change in health conditions of beneficiaries using such time-trend analysis tools.

Another relatively new set of tools includes the use of topological data analysis to identify new patterns of aberrant behavior. These have the potential to result in more accurate predictions. The main challenges of operating successful big data analytics programs may remain as the limited ability to integrate and manage unstructured data such as demographics or text data, data quality issues, and the potential shortage of analytics skills. These challenges can only be addressed in the long run through more training and investment.

KEY TAKEAWAYS

1. Unsupervised methods are used to explore claims data using unlabeled data and to reveal emerging fraud schemes.

2. Anomaly-based methods can be used for comparisons with peer groups to identify unusual activities.

3. Clustering methods can be used to group claims.

4. Bayesian hierarchical methods can help with both grouping as well as outlier detection.

5. Association type algorithms can reveal links among claims, providers, and patients.

6. Descriptive statistical methods can complement unsupervised methods while visualizing the output.

7. Direct fiscal impact, as well as potential improvements in fraud, waste, and abuse prevention, detection, and deterrence, should be considered while measuring the effectiveness of a statistical tool.

8. The deployment of rule-based methods can be improved by incorporating investigators and health care providers into the process.

9. Use of complementary data sources, consideration of changing legislation and health patterns, and frequent updating of rules can help decrease the ratio of false leads.

10. The choice of appropriate fraud analytics tool depends on the particular case and data.

11. Ease of use for data analysis and visualizations as well as data management are important considerations when choosing a statistical fraud detection tool.

ADDITIONAL RESOURCES

1. Data mining textbooks:

   a. James, G., Witten, D., Hastie, T., and Tibshirani, R. (2013). *An Introduction to Statistical Learning*. Springer. Available at www-bcf.usc.edu/~gareth/ISL/

   b. Hastie, T., Tibshirani, R., and Friedman, J. (2009). *The Elements of Statistical Learning: Data Mining, Inference, and Prediction*. Second Edition. Springer. https://web.stanford.edu/~hastie/Papers/ESLII.pdf

   c. Shmueli, G., Bruce, P.C., Yahav, I., Patel, N.R., and Lichtendahl Jr, K.C. (2017). *Data Mining for Business*

Analytics: Concepts, Techniques, and Applications. R. John Wiley & Sons. New York, NY.

2. Scholarly articles about use of unsupervised data mining algorithms for health care fraud detection

   a. Aral, K. D., Guvenir, H. A., Sabuncuoglu, I., and Akar, A. R. (2012). A prescription fraud detection model. *Computer Methods and Programs in Biomedicine*, 106(1):37–46.

   b. Ekin, T., Ieva, F., Ruggeri, F., and Soyer, R. (2013). Application of Bayesian methods in detection of health-care fraud. *Chemical Engineering Transactions*, 33:151–6.

   c. Ekin, T., Ieva, F., Ruggeri, F., and Soyer, R. (2018) Statistical medical fraud assessment: Exposition to an emerging field. *International Statistical Review*. 86(3), 379–402.

   d. Ekin, T., Ieva, F., Ruggeri, F., and Soyer, R. (2017). On the use of the concentration function in medical fraud assessment. *The American Statistician*. 71(3):236–41.

   e. Ekin, T., Lakomski, G., and Musal, R. (2015). An unsupervised Bayesian hierarchical method for medical fraud assessment. https://github.com/Prof-Greg/LDA.

   f. Iyengar, V. S., Hermiz, K. B., and Natarajan, R. (2014). Computer-aided auditing of prescription drug claims. *Health Care Management Science*, 17(3):203–14.

   g. Lin, C., Lin, C.-M., Li, S.-T., and Kuo, S.-C. (2008). Intelligent physician segmentation and management based on KDD approach. *Expert Systems with Applications*, 34(3):1963–73.

   h. Lu, F. and Boritz, J.E. (2005). Detecting fraud in health insurance data: Learning to model incomplete Benford's Law distributions. *In European Conference on Machine Learning*, 633–40. Springer.

i.   Major, J.A. and Riedinger, D.R. (2002). Efd: A hybrid knowledge/statistical-based system for the detection of fraud. *Journal of Risk and Insurance*, 69(3):309–24.

j.   Musal, R. (2010). Two models to investigate Medicare fraud within unsupervised databases. *Expert Systems with Applications*, 37(12):8628–33.

k.   Shan, Y., Murray, D.W., and Sutinen, A. (2009). Discovering inappropriate billings with local density based outlier detection method. In *Proceedings of the Eighth Australasian Data Mining Conference*, 101:93–8.

l.   Zafari, B. and Ekin, T. (2018). Topic modelling for medical prescription fraud and abuse detection. *Journal of Royal Statistical Society Series C*.

CHAPTER 6

Challenges, Opportunities, and Future Directions

OVERVIEW

Types of health care fraud and the methods used to address this crucial problem have been presented in detail so far. This book provides an overview of anti-fraud efforts using examples from U.S. health care programs. However, all other countries have to deal with similar issues.

For instance, in Australia, the fraud hotline received 1,116 Medicare-related tip-offs in a nine-month period in 2013–2014. These leads resulted in 12 convictions for a recovery of nearly half a million Australian dollars. Some experts claim the revealed fraud is just the tip of the iceberg and argue for the inefficiency of the mostly reactive fraud detection mechanisms (Scott and Branley, 2014).

It is still an ongoing debate if these cases are isolated or signs of large amounts of uncovered fraud. Some providers are found to be properly billing for services that are not provided and getting paid. For instance, a health care practitioner was billing properly with respect to payment rules and got paid almost A$4 million between 2006 and 2013 by Medicare. Only after an investigation, it was found out that he had fabricated more than 14,000 claims using the details of his patients without their knowledge. When he was finally caught after seven years, he already had frauded the system more than A$800,000. Are there more providers like him who are staying under the radar? We may never know for sure. That is the cruel truth about dealing with health care fraud.

What about England? It is reported that the National Health Services (NHS) could be losing up to £5.7 billion every year from its £100 billion budget to overpayments (Triggle, 2015). One example is the case of a dentist who was convicted of stealing £1.4 million from the NHS. She billed at least three times as much compared to similar dental practices. This was not because she was working very hard. She simply devised a scheme in which she was billing NHS the maximum amount for virtually every patient, alive or dead (!), once every three months. In addition, she was found to submit bogus claims for visits to nursing homes in further locations as far as Manchester, which is more than 320 kilometers (approximately 200 miles) away from London.

The situation is not estimated to be much different in developing countries. The lack of investigations and higher levels of corruption can be argued to help health care fraud stay buried under sand. For instance, an emerging fraudulent scheme in developing countries targets expats and tourists. Nowadays, more medical tourists are traveling in anticipation of lower hospital bills. In countries where there are relatively low levels of billing transparency and less regulation of the medical profession, hospitals may overtreat these patients and charge higher rates than normal.

In the United States, total Medicare savings were reported to be almost $17 billion in 2015. The prevention efforts including pre-payment reviews correspond to 84% of Medicare savings whereas recovery efforts have been reported to be 15%. The Fraud Prevention System (FPS) activities correspond to $604 million in savings in fraudulent payments being stopped, prevented, or identified. The implementation of analytical methods can be argued to be a success so far. In the U.S, current objectives include expansion of the implementation of statistical and analytical methods to all health care program integrity efforts and devising proper methods to assess their effectiveness. These require better communication and best practice sharing among federal, state, and private investigators.

There are countless ongoing efforts, opinions, and challenges associated with health care fraud assessment. It is unfair to suggest that all these can be summarized in a short book like this. The objective of this chapter is to provide further discussions and different perspectives on fraud control systems, with an emphasis on statistical tools. In addition to the prosecutors' or data analysts' point of view, this chapter also provides the physicians' point of view, as well as the legal aspects. A summary of recent and potential developments is also presented. Some interesting questions that are addressed are listed below:

- Who are the main shareholders in a health care fraud assessment framework?

- What are the different perspectives with regards to fraud control?

- What are the organizational issues about health care fraud assessment?

- Is change inevitable?

- Is data a blessing or a curse?

- How do you deal with the uncertainty associated with the outcomes of statistical methods?

- What are the challenges and opportunities going forward?

- What is next?

SHAREHOLDERS: PUTTING A FACE ON FRAUDSTERS AND VICTIMS

Health care systems have various shareholders at different capacities including insurers, providers, and patients. All these have responsibilities to control and report fraud. However, most do not own their role in the health care fraud control frameworks. Some of the providers and patients implicitly believe they do not owe anything to the system since they have never done anything wrong or committed any mistakes.

To start with, patients are not aware of their responsibilities within the fraud control framework. Obviously, it is their right to be provided health care. But in general, other than their own commitments such as the co-payment fee, patients do not check the billings that are made for their visits and may not care about the correctness of the transactions. They often lack the knowledge and incentives to self-audit these statements. To many, the transactions are between the government and providers. Hence, there is no reason to worry about the expenditure of someone else's money.

The lack of sympathy toward insurance companies and government health programs as potential victims of fraud also motivates that indifference. In a survey, one out of four Americans were found overstating the value of insurance claims to be acceptable, and more than 10% of them approved of submitting false claims for items that weren't lost or damaged (CNN, 2003). Health care fraud is seen as a victimless crime against the rich.

Insurance companies can be seen as overcharging. Therefore, some payback such as loaning a health insurance card to a family member not covered by the plan may not make people feel like fraudsters. Sharing what they have with people in need may feel nice. However, these small gestures add up and as small as they can be, they are illegal. Most of the time, insurers pass on such losses to those who pay premiums. So, the real victims become taxpayers and patients.

This is a regulation and policy issue as well as a behavioral issue. To start with, lack of coverage, high health care costs, and the emotional distress of being sick do not make people friendly to the system. The amount of paperwork and complexity of the system do not make providers' billing or patients' checking for their claims easy. These factors also challenge fraud investigators to figure out if a billing is simply a clerical mistake or an intentional fraudulent transaction. Therefore, fraudulent claims are mostly revealed after the fact, and the system tries to recover the payment, which is challenging on its own.

The ultimate goal of fraud control systems is preventing fraud and avoiding payments for fraudulent transactions in the first place. Fraud prevention is less flashy but is even more important than fraud detection. Therefore, continuous training with shareholders is very crucial. In order to avoid overpayments, identification of fraud at the point of claim submission is crucial. This can be achieved by prepayment and real-time fraud detection in which medical insights are incorporated. This requires a good understanding of simple clerical mistakes and suspicious claims submitted by potentially risky providers and patients. The more quickly these leads can be processed, the better it is. Training providers to prevent avoidable clerical mistakes without increasing their burden and time spent on billing can help facilitate fraud control. Providers should work within the fraud control system, and they should help investigators, rather than feeling that fraud control is an unnecessary burden.

Medical professionals are the most crucial part of any health care system. In reality, the fraudster providers make up a small portion. They abuse the public's well-deserved unconditional trust on providers. One would think that providers in general help investigators by helping to reveal such bad apples. When one of them has been found to be involved with any outright criminal fraud, providers should be the first to condemn that. However, even the professional associations rarely speak out against such conduct. They may not want to risk their subjectivity and decision-making being judged by a third party.

There is subjectivity involved with many clinical decisions, however in some apparent cases especially the ones that have adverse impacts on patients' health, the professional associations of physicians should be more vocal. The rotten apples in the system benefit from that ambiguity, and the immunity and protection that are afforded to their innocent peers. A few fraudulent providers may not have direct effects on other providers. However, their existence affects the overall system adversely.

This lack of cooperation makes the job of fraud examiners more challenging. When medical necessity and appropriateness is of concern, prosecutors generally may not pursue such cases, not risking the fight with doctors. The provider associations should ensure they provide rigorous training on billing practices and professional ethics similar to any other for-profit business. On the other hand, fraud examiners should consider the subjectivity and challenges that doctors face. All shareholders should keep in mind that minimizing overpayments and recovering losses will improve access to affordable and high-quality health care.

The Centers for Medicare & Medicaid Services (CMS) have been providing continuously updated training for the shareholders of the insurance programs both online and on site. The Zone Program Integrity Contractors (ZPICs) and Medicare Administrative Contractors (MACs) regularly engage in education and program integrity activities. Organizations such as the National Health Care Anti-Fraud Association (NHCAA) and the Association of Certified Fraud Examiners (ACFE) also help the government educate billing analysts, insurers, providers, and patients in the battle against fraud.

One party that has been relatively overlooked in this fight has been patients. The NHCAA advises every patient to protect their health insurance policy number and identification card carefully, to communicate with officials if suspecting fraud, to read the policy and benefits statements, and to be aware of kickback schemes (NHCAA, 2017). There is increasing emphasis on educating patients online through resources such as www.stopmedicarefraud.gov/.

CHALLENGES WITH PAYMENT AND FRAUD CONTROL SYSTEMS

Most payment systems are based on fee for service, which requires careful overseeing of billings. Identification of medically unnecessary and excessive billings is difficult prior to the payment. Alternative financial structures are introduced to change incentives for shareholders, aiming to decrease the cost and proportion of fraud. For instance, managed care organizations are reimbursed a fixed amount per patient per month, regardless of the level of service they provide. This can help control the costs to a certain extent but may result in a lower level or even denial of health care delivery. Fraudsters adapt to the new set of incentives quickly.

The unconditional trust in providers is also reflected in the payment systems. The current payment practice is not constructed to hold for-profit businesses in the health care industry fully accountable. These systems are based on the assumption that submitted claims are based on medically necessary and actually provided medical services for correct diagnoses. The claims are paid within 30 days as long as they are billed in line with the regulations. Fraudsters can fabricate claims as long as they process them correctly and as expected by the mostly automated claims-processing software. Even in the cases of human inspection, they may get away with fraud if they know the rules and thresholds of the system.

For instance, during post-payment utilization reviews, providers are given the right to support their billings by providing extra medical records and other relevant documentation. This level of audit focuses on medical appropriateness. If the documents match the claims, the provider will most likely pass the audit. If the provider presents inadequate or no documentation, then those particular claims may be reversed, and the payments may be adjusted.

Most often, such providers are assumed not to know the system well enough to provide proper documentation and are trained in how to correct their billing behaviors for the future. However, if a fraud perpetrator bills with respect to known billing patterns, it is very unlikely to be subject to such review. That is why statistical methods are very important to discover new patterns of fraud.

In addition to the standard audits and reviews, many insurance programs also employ special investigative units. A small number of former police officers and law enforcement officials would be in charge of a few grand investigations. Their performance is generally evaluated based on the number of processed cases and dollars recovered or settlements obtained. For instance, the Health Care Fraud Prevention and Enforcement Action Team (HEAT) is a joint effort by the Department of Health & Human Services (HHS) and the Department of Justice (DOJ) which has been very successful. The HEAT Medicare Fraud Strike Force found a $50 million scheme in which home health care companies paid kickbacks to patient recruiters in return for referring patients for home health and therapy services that were not medically necessary and were not provided (OIG, 2014).

Although these special investigations are very beneficial in recovering overpayments, they are mostly reactive. These audits are generally conducted in areas where the last big scandal was. Emerging fraud schemes especially tend to be outside the scope of such reviews. These investigations take a long time, so organized fraudsters generally may not be accessible by the end of the review.

ORGANIZATIONAL ISSUES: "NO NEWS IS GOOD NEWS!"

"No news is good news!" summarizes the sentiment that revealing fraud may create in an organization. The trade-off between timeliness and caution is tough to manage. The process management and fraud control teams have conflicting objectives. In a highly automated setting with an enormous number of claims to be processed, fraud control can easily become an afterthought. A fraud control team can be seen as an obstacle by the processing team and providers. Their extra caution can be perceived to prevent timely processing of submitted claims.

Such emphasis on processing accuracy and efficiency over verification creates a safe haven for fraudsters. Therefore, fraud control should not be the role of one department or a select number of investigators. Overall, fraud control teams are understaffed and under-resourced. It should be a joint effort and all shareholders should play their respective roles.

One of the current challenges is the isolation of fraud investigation teams and their lack of communication. For instance, it is not unusual that a provider whose license was revoked by the Medicaid of one state can be still eligible for payment by the Medicaid of another state. As a remedy, the federal government tries to coordinate the effort to share the Medicare exclusion list with the states. The CMS also works with the states to improve the anti-fraud culture within each Medicaid system. In that direction, the Medicaid Integrity Institute serves as the national Medicaid program integrity training center for states.

There is a trade-off between performance and standardization. However, even simple efforts such as having a centralized database of excluded providers and emerging fraud schemes would help the overall fraud control effort. To address such isolation of programs and improve communication, the United Program Integrity Contractor (UPIC) role is designed to operate under consolidated Medicare and Medicaid Program Integrity audit groups.

Fraud detection and conviction are expensive. Experts, auditors and data analysts need to work together across many departments. It is time-consuming to have such an integrated team that can quickly check claims before payment. Even if they are successful, it is disturbing to reveal fraud. You need to make a strong claim, potentially ruining the comfort and satisfaction of patients, hospitals, and providers. You may be risking penalties if you do not comply with regulatory requirements for timely payment. You better be right since health care fraud is a big accusation with real-world consequences. You may risk future business and prospective negotiations within the network. If you are found to be targeting innocent providers, there would be many adverse consequences. Therefore, convicting fraudsters is not an easy task. That is why some simply prefer to accept the cost of fraud as a cost of doing business.

EVOLUTION OF FRAUD AND ADAPTIVE FRAUDSTERS

"Change is the only constant in life."
"You could not really step twice into the same river."

These quotes are widely attributed to the Greek philosopher Heraclitus of Ephesus (535 BC–475 BC). How right he turned out to be, at least about Ephesus! Back then, Ephesus was established as a coastal town by the Aegean Sea. The city was home to the famous Temple of Artemis, one of the *Seven Wonders of the Ancient World*. Twenty-five centuries later, it is more than nine kilometers (approximately six miles) away from the shore, with only its foundations and sculptural fragments resisting time.

Heraclitus was on to something. Health care and fraud are not any different. Legislation changes. Health care treatments change. Insurance policies change. Diseases change. Diagnoses change. Health care is very dynamic. Fraud continuously evolves, and fraudsters are adaptive. What does this mean for fraud assessment?

Fraud detection is an ongoing fight. New fraud types are emerging. The evolving nature of fraud schemes is the biggest issue for supervised methods. The training data may not remain relevant after a while. Such supervised models need to be tuned continuously. Therefore, unsupervised methods are gaining ground. They can adapt to dynamic data more easily.

The CMS has a process that aims to manage this dynamic behavior. The existing analytical models are continuously reviewed and updated. The resources are used to address the highest priorities. For instance, within the FPS framework, the FPS Operations Board, which consists of senior program integrity leaders, proposes changes on the models and finalizes each release in certain periods. This is expected to prevent supervised algorithms from becoming obsolete after a while, updating the training data sets to address changing fraud patterns.

DIFFERENT SIDES OF THE COIN: DATA AS A BLESSING, DATA AS A CURSE

Data analysis has been very beneficial in generating meaningful leads by identifying fraudulent activities, revealing duplicate claims, detecting unnecessary medical services and assessing the probability of fraud within specific provider groups. The amount of data is increasing in size rapidly, with, for instance, nearly 45,000 new providers applying for Medicare enrollment every month. While this presents many opportunities, it is also challenging to manage. The health care system needs to conduct fraud control while incorporating physicians and patients with minimum disturbance to the system.

Lawmakers have seen recoveries from health care fraud as one of the easy ways to balance budgets. Therefore, contractors who are paid per the amount of recovery are presented as a viable option. Unfortunately, data analysis and statistical methods may be abused by some contractors.

A significant number of health care policymakers and providers have doubts about the frequent implementation of fraud analytics. For instance, one of the main complaints includes the leads of the fraud analytics methods generating too many false positives. Small providers who cannot carry the cost of the legal burden may go out of business after an audit, even if they are innocent.

In addition, many providers argue that aggressive use of health care fraud analytics affects their ability to treat customers the best way and shapes the type and amount of care patients receive. Doctors may not be able to recommend the best course of treatment for their patients at the risk of being flagged and investigated.

There is resistance from many providers about incentive-based recovery programs. Providers complain about the excessive checks and paperwork, and sometimes temporary limitations on payment. They argue that contractors who work on a percentage-based system tend to be more eager to take risks and investigate more leads that can turn out to be false alarms. These aggressive audits may put an extra burden on providers resulting in potential loss of time and resources.

This is a legitimate concern in a variety of cases. For instance, automated detection systems can red-flag claims that require radically different treatment methods and different prescriptions. It can be argued that these false positives may have adverse effects on both the providers and investigators. Some providers may have to come under extra burden of the legal process to show they are innocent. Even if they are found to be innocent, the negative publicity can adversely affect their practice. The resources of the government and the investigator would also be wasted. Overall, this highlights the importance of proper evaluation of the tool outputs. The supervised models need to be tuned properly and often enough, the rule repositories need to be updated, and the thresholds need to be chosen carefully. The contractors should be careful about their ethical responsibilities.

On the other hand, such incentives and legal changes have helped whistleblowers to be more courageous. Providing up to 25% of recovered losses motivates whistleblowers in addition to the legal requirements of reporting any unlawful activity within 60 days.

There are many critics who argue that the improvement is not fast enough and ask for more aggressive statistical fraud detection. It has been argued by Parente et al. (2012) that the CMS could have saved $18.1 billion annually in Medicare Part B, far more than has been reported. The Population Reference Bureau projects the senior population in the U.S. to more than double to 98 million by 2060. Therefore, the importance of fraud detection can be expected to increase. The CMS and Congress have been pushing for wider implementation of fraud analytics. The CMS is planning to expand and improve models to identify bad actors more quickly and more effectively.

LEGAL CONCERNS: EMBRACING UNCERTAINTY

What about ways to prevent abuse of statistical methods? The widespread use of statistical methods may raise a number of legal concerns with respect to scientific admissibility standards and statistical proof. A number of legal offices provide consultation to providers in their appeals challenging the health care audits especially in the case of extrapolations. They work with statistical experts to find potential problems in the work of contractors and challenge the validity of statistical methodology. They rightfully argue that even a few dollars in claim errors in the audit sample may result in an enormous amount of dollars in extrapolation.

Supervised data mining methods may be challenged since they have another level of subjectivity as their results are based on labeled data. However, it should be noted that data mining methods are not the final phase of the audit process. Such output generally serves as a tool to generate leads.

In their evaluation of the reliability of the expert testimony, scientific methodology, and evidence, the courts generally refer to Daubert's reliability requirement. The considerations include the acceptance of the theory by the scientific community, its peer review and publication history, testing history, rate of error, as well as its independence from the related litigation. Reference class and blue bus problems are raised as potential concerns (Issar, 2015). What are these issues? Let's make an effort to present brief explanations.

How do we choose the related variables that are used in a statistical method? Choice of a proper reference class in health care audits refers to identifying what is relevant among the enormous pool of health care information. It should be noted that many data mining methods include attribute and model selection algorithms. Some argue for the need for the transparency of such black box methods.

Whereas the blue bus problem refers to the assessment of a probability statement. For instance, when the DNA evidence suggests that there is 99% probability that the suspect is the murderer, how confident do you feel? At what point can the probability be judged to be high enough to assert certainty? As used in statistical sampling and overpayment estimation, is 95% high enough? At what point do the proportions of false positives and false negatives become too high?

Some of these are still open-ended questions. We, as a society, are still in the process of embracing and living with uncertainty.

Most contractors have been doing a very fine job for government; however, their activities should be overseen appropriately. For instance, a lack of oversight has created issues in Texas between the contractor, Texas Medicaid, and the federal government. The Office of the Texas Attorney General sued a company because it approved the payment of $1.1 billion in claims for orthodontic services, a "substantial percentage" of which were allegedly paid in violation of Medicaid policies. Meanwhile, the federal government argued that Texas did not ensure that prior authorization process was used to determine the medical necessity of orthodontic services. They asked for $133 million back for unnecessary Medicaid payments. In the investigation, that company was found to be in the wrong by having hired high school dropouts as dental pre-authorization specialists, instead of medical professionals.

Another issue is potential data falsification. In Texas, an internal fraud investigation revealed that some of the data analysis output was not reproducible. An employee was found to have chosen samples that would increase the recovery amount. He admitted to sample falsification and not reviewing the sample before approval. $16 million worth of recovery was reduced to $39,000 because of faulty extrapolation due to falsification of data and invalid statistical methodology.

Statistical methods have been very helpful to recover overpayments and improve patient health. As in many things in life, checks and balances are crucial. Modeling assumptions and data collection steps should be transparent. The accuracy of the output should be validated.

A TAKE ON FUTURE

Most evidence suggests that health care fraud, waste, and abuse will be a bigger concern going forward. Health care expenditures are increasing at an accelerating rate. Health care spending is projected to grow 1.3 percentage points faster than GDP per year over this period; as a result, the health share of GDP is expected to rise from 17.5% in 2014 to 20.1% by 2025. Therefore, statistical fraud assessment methods will be very important in the future as well. More and better data, better models, accurate evaluation metrics, better integration with audit processes, and incorporation of medical expertise will be the game changers to increase the effectiveness and use of such systems.

Statistical methods should complement medical prevention, detection and response efforts. While fraud detection involves identifying fraud as quickly as it has occurred, fraud prevention describes the measures to stop fraud from occurring in the first place. Therefore, creating an anti-fraud culture and improving internal compliance systems have long-term effects against fraud.

Fraud detection through statistical methods improves actual health care delivery and protects patients' lives. Integrated tools that let providers and doctors check patient history can help with prescription fraud and prevent potential prescription drug abuse. Expanding reviews for questionable drug prescribing beyond controlled substances to other commonly abused drugs is a crucial step to prevent abuse. In that direction, CMS launched a web-based tool to share information and actions against pharmacies deemed high risk. Statistical methods will continue to play a big role in these efforts to decrease drug abuse.

It is important to make patients a part of fraud prevention and detection schemes. Improving information security will be a major priority to help patients preserve their own information. With increasing levels of protection for whistleblowers, more people will step forward to report fraud. Ponemon Institute, a data security research firm, found that 50% of health care identify theft victims did not report errors on their statements at all. Even when patients do identify errors, it can be hard to know where to report them. There needs to be more training and easy-to-use tools that can help patients understand fraud, waste, and abuse, and report them. More data analytics can also be used for preventing potential health care data breaches. Data from public websites and log files can be used to spot abnormal server access patterns and perform security forensics.

Along those lines, current Medicare education campaigns emphasize that patients can help by:

- Never giving out their Medicare or Social Security number to anyone except their authorized provider

- Reporting any suspicious activities like being asked over the phone for their Medicare/Social Security number or banking information

- Checking their billing statements and reporting suspicious charges

- Using a calendar to track doctors' appointments and services helps quickly spot possible fraud and billing mistakes

- Checking their records by logging into mymedicare.gov

In addition, current engagement models that aim to engage patients in medical decision-making and their treatment will help the efforts against fraud prevention indirectly. Educated patients are generally more invested and capable of reporting potential irregularities associated with the billings for them. The variations between patients can be better measured using patient-generated health care data via wearables and social media.

Going forward, having a consensus about application of payment rules between payers and providers is very crucial. This can be achieved through continuous training and interaction, getting inputs from providers, and increasing the transparency of billings and respective audits. One of the main arguments explaining the inefficiencies of the U.S. health care system is that the payment systems are based on quantity. Providers being paid per procedure or prescription may incentivize profit-maximizing behavior. The main incentives should encourage keeping people healthy. Therefore, incorporating health outcomes to the evaluation of the billings may help increase the quality and efficiency of the health care systems.

The Fraud Prevention System (FPS) of the CMS is among the largest predictive analytics-based program of its kind in government. There are ongoing efforts to increase the rate of moving from a pay-and-chase to a prevention model of fraud detection. The next step would include sharing information with states about the expansion of analytical methods for their respective Medicaid programs. The CMS should provide the know-how to ensure data integrity. This collaboration would also strengthen the existing analytical models.

Such ongoing efforts include the Medicaid Integrity Program and the Medicare-Medicaid Data Match Program (Medi-Medi). Through Medi-Medi, the ZPICs collaborate with state Medicaid agencies to jointly investigate potential fraud and abuse. Integrated models that can combine the power of statistical models and expert opinion within a decision-making framework can be expected to be popular. Algorithm efficiency can be assessed within the decision support system. For instance, the consequences of false leads can be assessed better. Even a simple decision analysis setup would be useful for the evaluation of health care fraud detection tools and utilization of their output.

Electronic health care data have significantly improved efficiency and outcomes. The use of integrated data sources such as text data can be expected to be more important. However, data privacy and security of confidential data will be a bigger concern in the future. The emergence of frequent hacking can be expected to occur in the health care domain as well. Health care cybersecurity solutions and secure databases will be in higher demand as we move forward.

Recent advances fuel the argument for the need for purely automatic systems. However, the complexity and the nature of health systems may prevent that from happening in the very near future. The current focus has been on automated pre-payment analytics and proactive fraud detection. Proactive fraud assessment can become an integral part of the fraud prevention and detection frameworks, especially for cases with initial evidence. The contractors may be given more power to stop the payments in such cases despite the public outcry of providers. Statistical methods continue to be a powerful weapon in this fight because they allow organizations to combine, integrate, secure, and analyze large quantities of data. More advanced industry analytical platforms are proposed to handle various types of data and fraud patterns in real time effectively. Eventually, artificial intelligence can be embedded to mimic fraud expert opinion and medical insights.

Finally, a major pressing issue is the lack of statistical understanding among shareholders. Statistical theory can be used to provide an understanding of the uncertainty involved. However, a statistician is not an oracle. He/she cannot give a certain answer about future but can express uncertainty using probability statements. Such uncertainty in results may be difficult to communicate and is not understood in general.

Even when a new fraudulent network is discovered using a statistical tool with strong evidence, the fight is not over. It generally takes a long court battle and maybe eventually a settlement. In court cases, a judge would listen to statisticians from both sides and can be confused with the lack of certainty. Overall, increasing statistical literacy should be an important objective for the fight against health care fraud. This has been one of the main objectives of this book.

There is a not a new day without an algorithmic or technological development. We use statistics to understand, display, and improve the performance of systems. Statistics is on our side to help with summarizing the available information and deciphering uncertainty in our battle against health care fraud, waste, and abuse.

Health care fraud assessment is not a sprint, it is a marathon.

KEY TAKEAWAYS

1. Health care fraud is a global emerging concern that warrants more attention.

2. The aging population and increasing pressure of budget deficits will keep statistical methods as crucial components of health care fraud detection and prevention systems.

3. More training is necessary to make the shareholders of the medical systems aware of their roles and responsibilities in fraud control.

4. Fraud detection systems need to be adaptive to address the ability of fraudsters adjusting their schemes.

5. Fraud prevention should be the emphasis of fraud control frameworks.

6. Data analysts should be aware of limitations, strengths, and data compatibility of health care fraud detection methodologies.

7. The use of statistical methods may raise legal concerns with respect to scientific admissibility standards and statistical reasoning.

8. Integrated models that can combine the power of statistical models and expert opinion within a decision-making framework are in demand.

9. Health care fraud prevention is not a sprint, it is a marathon that requires adjustments and constant attention from all shareholders.

ADDITIONAL RESOURCES

1. Relevant resources for anti-fraud training:

   a. National Health Care Anti-Fraud Association (NHCAA): www.nhcaa.org

   b. Association of Certified Fraud Examiners (ACFE): www.acfe.com

   c. Help fight Medicare fraud: https://www.medicare.gov/forms-help-and-resources/report-fraud-and-abuse/fraud-and-abuse.html

   d. How to report Medicare fraud: https://www.medicare.gov/forms-help-and-resources/report-fraud-and-abuse/report-fraud/reporting-fraud.html

Bibliography

CMS (2014). Report to Congress: fraud prevention system second implementation year. The Centers for Medicare & Medicaid Services. https://www.stopmedicarefraud.gov/fraud-rtc06242014.pdf.

CMS (2016a). CMS research statistics data and systems. The Centers for Medicare & Medicaid Services. https://www.cms.gov/Research-Statistics-Data-and-Systems/Research-Statistics-Data-and-Systems.html.

CMS (2016b). The health care fraud and abuse control program protects consumers and taxpayers by combating health care fraud. The Centers for Medicare & Medicaid Services. https://www.cms.gov/Newsroom/MediaReleaseDatabase/Fact-sheets/2016-Fact-sheets-items/2016-02-26.html.

CNN (2003). 25% think insurance fraud is OK. CNN. http://money.cnn.com/2003/02/20/pf/insurance/insurance_fraud/.

Issar, N. (2015). More data mining for medical misrepresentation: Admissibility of statistical proof derived from predictive methods of detecting medical reimbursement fraud. *N. Ky. L. Rev.*, 42(2):341–74.

Joseph, L. and Rice, S. (2017). Texas feels health care's costly drain. Dallas News. https://www.dallasnews.com/business/economic-snapshot/2017/04/02/texas-feels-health-cares-costly-drain.

Leonard, K. (2016). For the first time, health care spending higher than social security. US News. https://www.usnews.com/news/articles/2016-01-25/health-care-programs-contribute-to-increasing-federal-deficit.

NHCAA (2017). The challenge of health care fraud. National Health Care Anti-Fraud Association. https://www.nhcaa.org/resources/health-care-anti-fraud-resources/the-challenge-of-health-care-fraud.aspx.

OECD (2015). Healthcare costs unsustainable in advanced economies without reform. The Organisation for Economic Co-operation and Development. http://www.oecd.org/health/healthcarecosts unsustainableinadvancedeconomieswithoutreform.htm.

OIG (2014). Annual report of the Departments of Health and Human Services and Justice. Office of Inspector General. https://oig.hhs.gov/publications/docs/hcfac/fy2014-hcfac.pdf.

OIG (2017). LEIE downloadable databases. Office of Inspector General. https://oig.hhs.gov/exclusions/exclusions_list.asp.

Parente, S.T., Schulte, B., Jost, A., Sullivan, T., and Klindworth, A. (2012). Assessment of predictive modeling for identifying fraud within the Medicare program. *Health Management, Policy and Innovation*, 1:8–37.

Petroff, A. (2017). U.S. health care admin costs are double the average. CNN. http://money.cnn.com/2017/01/11/news/economy/healthcare-administrative-costs-us-obamacare/index.html.

Ruiz, R. (2017). U.S. charges 412, including doctors, in $1.3 billion health fraud. *The New York Times*. https://www.nytimes.com/2017/07/13/us/politics/health-care-fraud.html.

Scott, S. and Branley, A. (2014). Australian Medicare fraud revealed in new figures, 1,116 tipoffs so far this financial year. ABC Australia. http://www.abc.net.au/news/2014-03-06/australians-defrauding-medicare-hundreds-of-thousands-of-dollars/5302584.

Sparrow, M.K. (2000). *License to steal: How fraud bleeds America's health care system*. Basic Books, New York, NY.

Triggle, N. (2015). Fraud could be costing NHS in England 5.7bn a year, says report. BBC. http://www.bbc.com/news/health-34326934.

Weaver, J. (2017). Bribes to low-paid state worker key to $1 billion Miami Medicare fraud case, prosecutors say. *The Miami Herald*. http://www.miamiherald.com/news/local/article164232522.html#storylink=cpy.

Yin, S. (2017). Hospitals are clogged with patients struggling with opioids. *The New York Times*. https://www.nytimes.com/2017/08/21/health/hospitals-opioid-epidemic-patients.html.

Index

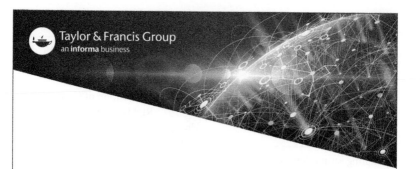

Printed and bound by CPI Group (UK) Ltd, Croydon, CR0 4YY

21/10/2024

01777044-0002